STEAM Education

Myint Swe Khine • Shaljan Areepattamannil
Editors

STEAM Education

Theory and Practice

 Springer

Editors
Myint Swe Khine
Emirates College for Advanced Education
Abu Dhabi, United Arab Emirates

Shaljan Areepattamannil
Emirates College for Advanced Education
Abu Dhabi, United Arab Emirates

ISBN 978-3-030-04002-4   ISBN 978-3-030-04003-1   (eBook)
https://doi.org/10.1007/978-3-030-04003-1

Library of Congress Control Number: 2018968253

© Springer Nature Switzerland AG 2019
This work is subject to copyright. All rights are reserved by the Publisher, whether the whole or part of the material is concerned, specifically the rights of translation, reprinting, reuse of illustrations, recitation, broadcasting, reproduction on microfilms or in any other physical way, and transmission or information storage and retrieval, electronic adaptation, computer software, or by similar or dissimilar methodology now known or hereafter developed.
The use of general descriptive names, registered names, trademarks, service marks, etc. in this publication does not imply, even in the absence of a specific statement, that such names are exempt from the relevant protective laws and regulations and therefore free for general use.
The publisher, the authors, and the editors are safe to assume that the advice and information in this book are believed to be true and accurate at the date of publication. Neither the publisher nor the authors or the editors give a warranty, express or implied, with respect to the material contained herein or for any errors or omissions that may have been made. The publisher remains neutral with regard to jurisdictional claims in published maps and institutional affiliations.

This Springer imprint is published by the registered company Springer Nature Switzerland AG.
The registered company address is: Gewerbestrasse 11, 6330 Cham, Switzerland

Introduction

The twenty-first century has been a unique era of technological advances and spread of globalization that surpassed the events in previous decades. Educational planners around the world endeavour and make every effort to educate the next generation of students to become technology literate and take interest in subjects such as science, technology, engineering, and mathematics in the face of increased economic competitions. However, these subjects are not to be taught separately but to be integrated into a cohesive interdisciplinary approach. This approach connects among discrete disciplines and converges into an entity, known as STEM. According to Tsupros, Kohler, and Hallinen (2009), "STEM education is an interdisciplinary approach to learning where rigorous academic concepts are coupled with real-world lessons as students apply science, technology, engineering, and mathematics in contexts that make connections between school, community, work, and the global enterprise enabling the development of STEM literacy and with it the ability to complete in the new economy".

Along with the progress in STEM curriculum, educators further advocate that in order to function well in the future society, the young generation must be equipped with the twenty-first-century skills that include creativity, innovation, and entrepreneurship. There is a growing debate among educators that "arts" should be integrated in the STEM curriculum to spur much needed creativity and innovation (Guyotte, Sochacka, Costantino, Walther, & Kellam, 2014). Adding arts and design to the equation will transform STEM into STEAM (Liao, 2016). Ge, Ifenthaler, and Spector (2015) refer to STEAM as "the inclusion of the liberal arts and humanities in STEM education; some STEAM conceptions simply use the 'A' to indicate a fifth discipline area—namely, arts and humanities". The notion of STEAM (science, technology, engineering, arts, and mathematics) is an emerging discipline unique in its desire to provide a well-rounded approach to education (Rolling, 2016).

Henriksen (2017) rightly pointed out that viewing STEAM as solely about arts integration is problematic since many science teachers may lack artistic training. There is an urgent need to explore on how arts can be integrated meaningfully in the STEM. While the information about STEM education is abundant in the literature, theory building, best practices, and practical applications in this new interdisciplinary area are sparse. The primary objective of this book is to fill this gap. The chapters in this volume examine STEAM in a variety of settings, from elementary and middle schools to higher education. Readers will benefit from the experience, novel approaches, and proven strategies shared by the pioneers and curriculum innovators in sustaining successful STEAM initiatives in schools and beyond.

The book begins with the chapter by Stroud and Baines (Chap. 1). By employing the tenets of inquiry-based learning and exploring the possibilities of inquiry-based learning, the authors postulate that the flexibility in choosing the appropriate methodologies is indispensable for correctly responding to important scientific, mathematical, technological, and engineering questions. A thorough examination of the philosophical and practical differences between scientific inquiry and the engineering design process is provided to better explicate the inquiry-based approach to learning. The authors, using examples from the applications of interactive notebooks and photo narratives, posit that integrating images and the arts into inquiry-based approaches to science and engineering may significantly enhance the quality of student learning. Stroud and Baines also describe the ways in which interactive notebooks and photo narratives necessitate students to use writing as a way of capturing real-time experience. The authors argue that "integrating the arts, requiring students to think and write, and stimulating students' creative impulses can promptly dispel any notions of ennui and help immeasurably in transforming STEM into a thoroughly contemporary STEAM".

In Chap. 2, Bush and Cook share their journey of working with in-service elementary teachers in school-level professional learning communities as they develop, refine, and implement three problem-based STEAM investigations (i.e., designing a prosthetic arm for a kindergartener, a palaeontology investigation, and a closer look at the arts within roller coaster engineering) in their classrooms with a critical focus on grade-level mathematics and science, as defined by the Common Core State Standards and the Next Generation Science Standards. The authors documented that the incorporation of "empathy" was one of the most significant and powerful aspects of integrated STEAM that separates it from integrated STEM. Hence, they posit that "STEAM investigations need not be planted or contrived real-life scenarios—instead they can be actual authentic scenarios facing the community in which students live". In addition to the important lessons learned from the implementation of these three problem-based STEAM investigations, Bush and Cook list a number of useful tips for successfully implementing high-quality STEAM instruction.

Liao, in Chap. 3, employing content analysis, examines the existing curricular approaches to STEAM education and describes the relationships between them. The author put forward a "STEAM map" and argued that it should "serve as a way to locate the goals and approaches of current STEAM practices and as a basis for art educators and other stakeholders to envision advancing into other areas".

In the next chapter, Henriksen, Mehta, and Mehta introduce design-thinking aspects in the curriculum. In their chapter, the authors propose a framework—design thinking—for integrating STEM and the arts. The authors describe the interconnections between design, design thinking, and STEAM and elucidate the ways in which classroom teachers can use design-thinking practices to redesign curriculum to

Introduction   vii

transition from STEM to STEAM education. In Chap. 5, Miller, using a constructionist approach delivered via instructional methods incorporating 2D and 3D technologies during STEM instructional activities within a creative space, examines the effects of makerspace professional development activities on elementary and middle school educators' perceptions of integrating technologies with STEM. The study demonstrated a significant increase in elementary and middle school educators' self-reported competence in technology integration, confidence levels towards integrating World Wide Web, emerging technologies for student learning, teacher professional development, and attitudes towards math, technology, science, and STEM careers.

In Chap. 6, Marmon investigates the impact that creativity and the arts have on traditional STEM courses. The author also examines the attitudes of the students towards classroom-based STEAM activities. Mehta, Keenan, Henriksen, and Mishra in Chap. 7 propose a threefold, iterative framework that may help integrate aesthetics into STEM learning. The authors also provide several examples for using the rhetoric aesthetics as a means to guide teacher professional development for STEM teachers. In Chap. 8, Quigley, Herro, and Baker examine how STEAM teaching practices are enacted in a variety of educational settings to better understand the teachers' STEAM curriculum implementation strategies. In the final chapter, Sundquist investigates the impact that the diversity of students' academic disciplines has on learning in collaborative group projects in a STEAM course.

It is hoped that this book is an invaluable resource for teachers and teacher trainers, university faculty, researchers, and school administrators. It will also be of interest to science, mathematics, engineering, computer science, information technology, arts and design, and technology teachers.

References

Ge, X., Ifenthaler, D., & Spector, J. M. (Eds.). (2015). *Emerging technologies for STEAM education: Full STEAM ahead.* Dordrecht, The Netherlands: Springer.

Guyotte, K. W., Sochacka, N. W., Costantino, T. E., Walther, J., & Kellam, N. N. (2014). STEAM as social practice: Cultivating creativity in transdisciplinary spaces. *Art Education, 67*(6), 12–19.

Henriksen, D. (2017). Creating STEAM with design thinking: Beyond STEM and arts integration. *The STEAM Journal, 3*(1), 11.

Liao, C. (2016). From interdisciplinary to transdisciplinary: An arts-integrated approach to STEAM education. *Art Education, 69*(6), 44–49.

Rolling Jr., J. H. (2016). Reinventing the STEAM engine for art+ design education. *Art Education, 69*(4), 4–7.

Tsupros, N., Kohler, R., & Hallinen, J. (2009). *STEM education: A project to identify the missing components.* Pittsburgh, PA: Intermediate Unit 1 and Carnegie Mellon.

Contents

1 Inquiry, Investigative Processes, Art, and Writing in STEAM 1
Adam Stroud and Lawrence Baines

2 Structuring STEAM Inquiries: Lessons Learned from Practice 19
Sarah B. Bush and Kristin L. Cook

**3 Creating a STEAM Map: A Content Analysis
of Visual Art Practices in STEAM Education** 37
Christine Liao

**4 Design Thinking Gives STEAM to Teaching:
A Framework That Breaks Disciplinary Boundaries** 57
Danah Henriksen, Rohit Mehta, and Swati Mehta

**5 Investigating the Impact of a Community Makers'
Guild Training Program on Elementary and Middle
School Educator Perceptions of STEM (Science, Technology,
Engineering, and Mathematics)** 79
Jennifer Miller-Ray

**6 The Emergence of the Creativity in STEM:
Fostering an Alternative Approach for Science,
Technology, Engineering, and Mathematics
Instruction Through the Use of the Arts** 101
Michael Marmon

**7 Developing a Rhetoric of Aesthetics: The (Often)
Forgotten Link Between Art and STEM** 117
Rohit Mehta, Sarah Keenan, Danah Henriksen, and Punya Mishra

**8 Moving Toward Transdisciplinary Instruction:
A Longitudinal Examination of STEAM Teaching Practices** 143
Cassie F. Quigley, Dani Herro, and Abigail Baker

9 Multidisciplinary Group Composition in the STEAM Classroom 165
John D. Sundquist

Chapter 1
Inquiry, Investigative Processes, Art, and Writing in STEAM

Adam Stroud and Lawrence Baines

Introduction

Science states meanings; art expresses them. John Dewey (1934, p. 84)

Modern educational practices tend toward integrating domains that were previously considered distinct and separate. In recent years, the term *STEM* has come to be affiliated with the fields of science, technology, engineering, and mathematics education. However, once STEM was established as a handy way of referencing these four fields in a concise acronym, scholars urged a further expansion to include the arts—and thus was born the term *STEAM* (Robelen, 2011).

Traditional views of STEM education emphasize theoretical understandings of solutions to real-world problems. However, within STEM education, the arts have always had a critical though unheralded place. It always has been de rigueur for scientists and engineers to construct models and to communicate conceptual understandings through diagrams, sketches, tables, and other modes of representation. The arts have a way of capturing the essence of an endeavor, reframing experience, and transforming perceptions. As Maxine Greene (1995) noted, "The arts provide new perspectives on the lived world" (p. 4).

Many of the most revered figures in the history of the world earned their reputations through contributions to both the sciences and art. It is well known that among Da Vinci's inventions can be counted the parachute, a robotic knight, a flying machine, a revolving bridge, and protective underwater clothing (Capra, 2008). Benjamin Franklin, Nikola Tesla, Albert Einstein, and John James Audubon are other luminaries who fit the artist-scientist archetype (Maeda, 2013). Perhaps less often acknowledged is the fact that these individuals were equally celebrated for their prodigious output of writings.

A. Stroud · L. Baines (✉)
University of Oklahoma, Norman, OK, USA
e-mail: lbaines@ou.edu

© Springer Nature Switzerland AG 2019
M. S. Khine, S. Areepattamannil (eds.), *STEAM Education*,
https://doi.org/10.1007/978-3-030-04003-1_1

In this vein, it seems useful to view the implementation of inquiry-based learning through a theoretical framework that supports the renegotiation of STEAM education. Students' perceptions of STEM as dull and boring (something that can be confounding and upsetting to teachers) can be transformed into extraordinary through judicious use of the A in STEAM and working in as much writing as possible.

Scientific investigations are typically bifurcated into two major methods: scientific inquiry and the engineering design process. Both methods attempt to construct knowledge in light of new understandings, and both methods attempt to communicate evidence-based claims to the rest of the world. As with art, at its core, scientific investigations are predicated upon curiosity and creativity.

The flexibility to choose the correct methodologies is essential for answering scientific, mathematical, technological, and engineering questions, so the first part of this chapter focuses upon the principles and possibilities of inquiry-based learning. Next, the chapter investigates the philosophical and practical differences between scientific inquiry and the engineering design process. Finally, the chapter concludes with specific suggestions for using the arts as prompts for writing with the goal of improving the quality of learning. Two activities that purposefully use images and writing and promote creativity—interactive notebooks and photo narratives—are presented in detail.

Dewey and the Importance of Experience

Traditionally, educational settings placed students in contexts that were situated in the classroom unconnected to the contexts of the content. For example, students might study the contamination of water by reading a textbook and examining photos rather than through active testing and analysis of water quality by the shore of an endangered river, led by an expert in environmental pollution. Dewey established that education needed a new philosophy of theory that used experience to support the ways in which learners construct knowledge. Emphasis was placed on the joining of education and experience. "The fundamental unity of the newer philosophy is found in the idea that there is an intimate and necessary relation between the processes of actual experience and education" (Dewey, 1938, p. 6).

In this way, learners engage with topics pertaining to content areas situated within contexts that support the construction of knowledge. The transition, proposed by Dewey, offered a joining of two very important constants: (1) the student and (2) lived experiences. As Schon (1992) notes, "Dewey treats human inquiry as continuous with the biological transaction between the organism and environment, hoping in this way to establish an objective basis for describing both what is problematic about problematic situations and what is determinate about their resolution" (p. 122).

Dewey continued to work on establishing an order of operations in which experience could be made replicable. If experiences are to be utilized within the classroom,

1 Inquiry, Investigative Processes, Art, and Writing in STEAM

then educators needed a practical method of implementing the theoretical constructs of inquiry. Dewey (1933) developed a procedure to support the construction of knowledge within a particular experience:

1. Observation of surrounding conditions
2. Knowledge of what has happened in similar situations in the past
3. Judgment which puts together what is observed to see what they signify

Although not expressly developed for science, Dewey emphasized the systematic collection of empirical evidence, gathered through the use of the senses, and the formulation of inferences based on those sensory observations. Since Dewey, Bruner (1961) offered an explanation of inquiry as the process of rearranging or transforming evidence in such a way that enables going beyond the evidence to forge additional new insights. Developing skills such as making observations and creating inferences can encourage learners to construct new knowledge and to ponder novel solutions. Because experience is the zeitgeist for learning, the quality of an experience is absolutely critical.

Piaget and the Learning Cycle

Jean Piaget sought to understand the process in which an individual comes to create knowledge based on experience. "Intelligence, viewed as a whole, takes the form of a structuring which impresses certain patterns on the interaction between the subject or subjects and near or distant surrounding objects" (Piaget, 1960, p. 167). Through his research, Piaget began to understand the investigative process as one in which individuals constantly interact with the environment. Piaget postulated that the process of knowledge creation was, in actuality, a series of recursive intuitions. Individuals move through a continual mental functioning process beginning with assimilation, transitioning to states of disequilibrium and accommodation, and ending with the integration of new knowledge (Piaget, 1960).

Piaget's research provided insight into the ways learners construct knowledge. By purposefully placing students in situations where the content to be learned is just beyond their cognitive reach, they enter into a state of disequilibrium. Bybee (1982) explains the process as an ongoing negotiation between disequilibrium and accommodation.

> The teacher can expose the student to problem situations that are slightly beyond their level. In doing so, the student will experience conflict and dissonance while applying the current level of cognitive structures to problematic situations at a higher level; thus disequilibration and equilibration. (p. 201)

Piaget asserted that by placing students in situations of cognitive dissonance and offering opportunities to equilibrate, students could build fully elaborated schemas of thought. Learning through inquiry can prompt scrutiny of the way knowledge becomes manifest and can open the possibilities for greater understanding. In this manner, knowledge loses the default assumption that is always absolute and eternal. Instead, claims to knowledge become tentative, susceptible to revision. Continual negotiation and reaffirmation are required for knowledge to be accepted as genuine (Massiala, 1969).

Defining Roles for the Teacher and Student

As the learner plays an active role in investigating phenomena, the teacher is there to provide direction, but not necessarily solutions. Karplus and Their (1967) saw the teacher/student relationship as a process of discovery, permitting students to learn how to learn so that new concepts could be formed. Establishing the roles for both the teacher and learner is critical to supporting how investigations are to be carried out. However, roles can be augmented to support the varying nature of inquiry. Depending on the needs of the learner and the objective of the investigation, there is a need to renegotiate the levels of responsibility placed on the teacher and the student.

Traditional methods of instruction differ from that of an inquiry-oriented instruction in that the responsibility for learning is placed upon the learner and wholly dependent upon their active participation. Aulls and Shore (2008) in Fig. 1.1 offer an organizational representation of the various responsibilities of students and teachers based on different kinds of instruction. As a teacher, it is useful to consider instructional strategies in light of desired goals, objectives, and aspirations.

Although inquiry recreates the student/teacher dynamic as far as redefining roles, it is important to consider the level to which students are currently capable of taking on these roles. Hawkins and Pea (1987) saw inquiry as having the potential to keep knowledge alive and to help students claim knowledge as their own. However, in reference to managing how students navigate the learning cycle, a gradual movement from direct instruction to inquiry learning might be appropriate. In teaching any curricular content, the needs and dispositions of students must be considered along with the appropriate instructional design.

For example, answering the questions "Why does this taste sour?" and "How can I make this taste sour?" calls for two very different approaches. Educators often begin with questions and ask students to conduct investigations that do not allow for the accumulation of evidence to support a potential solution. Well-defined problems tend to engender astute answers, while poorly defined problems cause confusion (Wheatley, 1998). It is the responsibility of the teacher to establish the environment under which a question is examined (Dewey, 1938). In other words, investigations should strategically place the learner in a position to connect new findings to prior knowledge as a matter of course.

1 Inquiry, Investigative Processes, Art, and Writing in STEAM

Kind of Instruction	Teacher Role	Student Role	Learning Achieved
Direct Instruction	-Present Content -Maintain Control	Limited Role	-Memory and Recall
Reciprocal Teaching	-Model -Facilitator	Group Member	-Learning how to learn -Prediction -Clarification/Summary
Discovery Learning	-Planner -Manager -Resource Person -Audience	-Expert -Hypothesis Generator -Data Collector/Analyzer -Author/Presenter	-Generative -Conceptual/Strategic -Generate theory inductively
Inquiry Learning	-Encourager -Organizer -Guidance -Evaluator	-Problem Solver -Observer -Investigator -Author -Debater	-Learning how to inquire -Conceptual Learning -Generate theory both inductively and deductive -Learning promotes understanding

Fig. 1.1 Organizational representation of different kinds of instruction (From Aulls and Shore, 2008)

Any investigation should support the learner during each stage of the learning cycle. If attention is not placed on the individual learner during an investigation, then the knowledge that is created can be inaccurate or imprecise. At all times, educators must be diligent to detect and help students become aware of misconceptions and address them on the spot. Students, who hold misconceptions during the beginning of inquiry and incorrectly equilibrate, may retain the misconception for some time (Aulls & Shore, 2008).

Methods of Investigation: Scientific Inquiry and the Engineering Design Process

Traditionally, educators may have viewed science instruction in the form of a single lesson experienced wholly within the predetermined time allotment for a class. One of the essential characteristics of inquiry is to allow the nature of the experience to dictate the time needed for an investigation. A well-designed lesson includes a detailed description of a learning experience that may be the focus for 1 day or several days (Hammerman, 2006). The 5E model of instruction seeks to develop continued opportunity for knowledge construction while assessing students as they move through the learning cycle. Hammerman (2006) gives examples of tasks seen in each phase (see Fig. 1.2).

Scientific practices may require the utilization of different methods to achieve desired goals. Thus, a student might approach a question using the structures of scientific inquiry, the processes of engineering design, or both. Such direct involvement gives students an appreciation of the wide range of methodologies that can be used both inside and outside the classroom to investigate, model, and explain the world around them.

Although each method of investigation is particularly suited to accomplish different sets of objectives, there exist commonalities between the two, including modeling, developing explanations, engaging in critical discussion, and formulating detailed evaluations (National Research Council, 2012).

The nature of thought represents a continual flow of assimilation and accommodation. Below is a diagram (Fig. 1.3) representing the three spheres of activity for scientists and engineers as presented by the National Research Council (2012).

Represented within each phase are three spheres of activity: investigating, evaluating, and developing explanations and solutions. As noted by the National Research Council (2012), "In reality, scientists and engineers move, fluidly and iteratively, back and forth among these three spheres of activity, and they conduct activities that might involve two or even all three modes at once" (p. 46).

In considering the needs of diverse student populations, "The Real World" may have a different look or a different sheen based upon the individual engaged in the

Engage:	Use a discrepant even to raise questions to see what students know.
Exploration:	Identify an inquiry question. Provide or collaboratively produce an action plan.
Explanation:	Use student generated data to support new learning. Build on prior knowledge. Identify and correct misconceptions or discrepancies.
Elaboration:	Ask new questions and engage in further investigation to extend learning.
Evaluation	Use assessment strategies to show evidence of learning throughout experience.

Fig. 1.2 The five Es (Hammerman, 2006)

1 Inquiry, Investigative Processes, Art, and Writing in STEAM

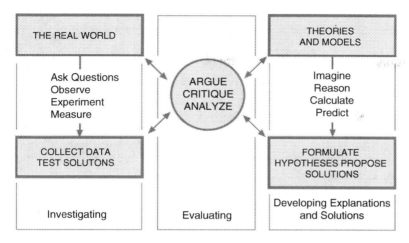

Fig. 1.3 A framework for K–12 science education (National Research Council p. 45)

experience. In this regard, it is important to understand that the variety of tools available for scientific investigations can support individual predilections, allow for new perspectives, and enable fresh insights.

Engineering design places emphasis on questions such as "What can be developed?". Engineering design often goes beyond investigation and into the realm of testing potential solutions through the building and testing of physical or mathematical models and prototypes, which provide data that could otherwise not be collected.

Through this systematic testing and evaluation, the learner is placed in the position of having to produce evidence-based claims for a presentable solution. Although students do the majority of the work during inquiry-based learning, the engineering design process relies heavily on the expertise of the teacher to direct both the content and procedural steps necessary to carry out an engineering investigation.

Successful implementation of the engineering design process relies on a different set of skills than scientific inquiry. Although students are still ultimately working toward generating a solution and communicating ideas to others, certain procedures must be followed. The National Research Council (2012) identified seven key components in the engineering design process:

1. Engineering begins with a problem that needs to be solved.
2. Engineering requires the use of models and simulations to find strengths and weaknesses.
3. Investigations allow students to gain data and to test designs.
4. Empirical evidence is required for claims to be made.
5. Models are based on knowledge of the material world.
6. Reasoning and argumentation are essential.
7. Improved technologies cannot be produced if ideas are not communicated clearly and persuasively.

Developing a solution through definition and optimization is a highly recursive process. If, at any point, the inquiry-based learning experience focusing on the engineering design process neglects one or more of these key components, then the process is potentially corrupt and incomplete. Defining problems and developing solutions subject to constant evaluation constitute the meat and potatoes of the engineering design process.

Before the engineering design process can take place, it is possible that another set of questions must be answered. In the case of early understandings of HIV/AIDS, groups of people were presented with the phenomena of HIV/AIDS, though it was little understood at the time. In this regard, two questions were instrumental in moving forward with the research:

1. "What is HIV?"
2. "What can be done about HIV?"

Of course, the most effective way to attempt to answer the question "What is HIV?" is through scientific inquiry. However, the most effective way to create an effective response to the question "What can be done about HIV?" is through the engineering design process.

Although K–12 learners may never be tasked with the monumental task of identifying a new infectious disease or developing state-of-the-art pharmaceuticals such as nucleoside reverse transcriptase inhibitors (NRTIs), it might be beneficial for students to learn how to approach such problems in an orderly, systematic, and scientific manner. Having students develop theory-based models and argue them based on evidence from observations in order to explain and demonstrate their knowledge to others would seem to be a good place to start.

Inquiry is based on the student's ability to make connections to the phenomena that is being observed. Although this is similar to how the engineering design process is carried out, one of the major differences is that there may or may not be an applicable solution created. As Schon (1992) stated, "Inquiry learning is concerned with solving problems but it does not require solutions to problems" (p. 181).

The types of questions that are presented within science inquiry require students to construct a different kind of knowledge. An investigation focused on engineering processes may ask: "What can be done to address a particular human need?" On the other hand, a science-based inquiry may ask: "Why does it happen?"

The National Research Council (2012) describes the practices in science inquiry as follows:

1. Formulating empirically answerable questions to investigate what is known and yet to be answered.
2. Predictions enable the creation of a model to represent knowledge.
3. Investigations are conducted both in the field and laboratory.
4. Investigations require controls, independent, and dependent variables.
5. Meaning is derived from data.
6. Reasoning and argumentation are necessary for the construction of theories that provide explanatory accounts of the world.

7. Without communicable ideas, science cannot advance.
8. Science inquiry and engineering design may look very similar in the process skills necessary to carry out an investigation. However, it is imperative to focus on the questions that guide the investigation and the overarching goals of science inquiry, which may or may not result in an applicable solution.

Student responsibilities increase as the classroom moves toward a more inquiry-based approach to learning. The importance of a positive and productive teacher/student rapport in the sciences is critical; thus, during any investigation, there must be ongoing, interactive discourse. However, when viewed through inquiry-based learning, natural phenomena routinely place students in a place of disequilibrium as misconceptions are plainly exposed. This can lead to low self-efficacy in learners. Jaladanki and Bhattacharya (2015) suggest that teachers consider interactive notebooks to increase student self-efficacy and to open up an ongoing dialogue between teacher and student. Interactive notebooks can provide students with a safe space to ruminate about class activities, while they offer teachers a genuine, unexpurgated gauge of student understanding and current dispositions.

Interactive Notebooks

Depending on the amount of information that the course will cover, or the nature of the material that will be kept within the notebook, a simple 1-subject, 100-page, spiral notebook can serve as the foundation for an interactive notebook. To begin the initial setup, students need copies of the rubric that will be used for assessment. Having students tape the rubric into the inside cover of the notebook can offer an easily accessible reference to what is being assessed. A table of contents is also needed to allow for both the student and teacher to access information. Figure 1.4 is an image of a student's front matter from an interactive notebook.

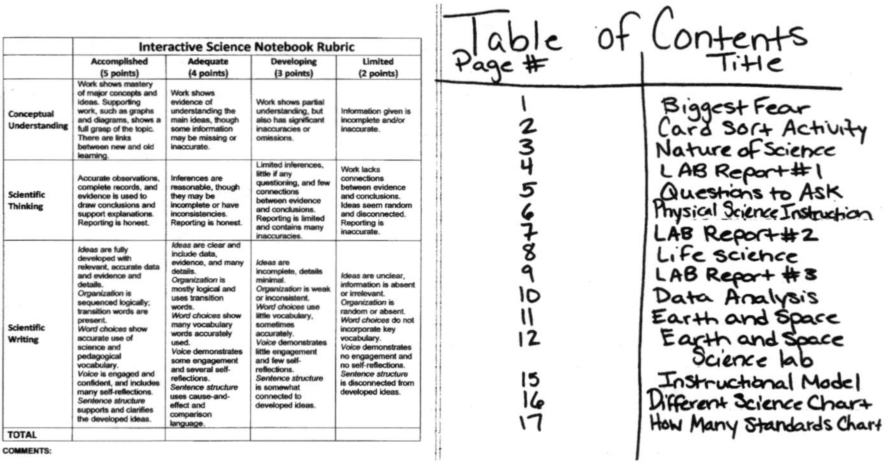

	Interactive Science Notebook Rubric			
	Accomplished (5 points)	Adequate (4 points)	Developing (3 points)	Limited (2 points)
Conceptual Understanding	Work shows mastery of major concepts and ideas. Supporting work, such as graphs and diagrams, shows a full grasp of the topic. There are links between new and old learning.	Work shows evidence of understanding the main ideas, though some information may be missing or inaccurate.	Work shows partial understanding, but also has significant inaccuracies or omissions.	Information given is incomplete and/or inaccurate.
Scientific Thinking	Accurate observations, complete records, and evidence is used to draw conclusions and support explanations. Reporting is honest.	Inferences are reasonable, though they may be incomplete or have inconsistencies. Reporting is honest.	Limited inferences, little if any questioning, and few connections between evidence and conclusions. Reporting is limited and contains many inaccuracies.	Work lacks connections between evidence and conclusions. Ideas seem random and disconnected. Reporting is inaccurate.
Scientific Writing	*Ideas* are fully developed with relevant, accurate data and evidence and details. *Organization* is sequenced logically; transition words are present. *Word choices* show accurate use of science and pedagogical vocabulary. *Voice* is engaged and confident, and includes many self-reflections. *Sentence structure* supports and clarifies the developed ideas.	*Ideas* are clear and include data, evidence, and many details. *Organization* is mostly logical and uses transition words. *Word choices* show many vocabulary words accurately used. *Voice* demonstrates some engagement and several self-reflections. *Sentence structure* uses cause-and-effect and comparison language.	*Ideas* are incomplete, details minimal. *Organization* is weak or inconsistent. *Word choices* use little vocabulary, sometimes accurately. *Voice* demonstrates little engagement and few self-reflections. *Sentence structure* is somewhat connected to developed ideas.	*Ideas* are unclear, information is absent or irrelevant. *Organization* is random or absent. *Word choices* do not incorporate key vocabulary. *Voice* demonstrates no engagement and no self-reflections. *Sentence structure* is disconnected from developed ideas.
TOTAL				

COMMENTS:

Table of Contents

Page #	Title
1	Biggest Fear
2	Card Sort Activity
3	Nature of Science
4	LAB Report #1
5	Questions to Ask
6	Physical Science Instruction
7	LAB Report #2
8	Life science
9	LAB Report #3
10	Data Analysis
11	Earth and Space
12	Earth and Space Science lab
15	Instructional Model
16	Different Science Chart
17	How Many Standards Chart

Fig. 1.4 Front matter of a student's interactive notebook

The interactive notebook should be structured in such a way as to allow students to address standards, develop writing skills, create drawings, make connections, and explore possibilities (Chesbro, 2006). In addition to these advantages, the interactive notebook also serves as documentation of discoveries, epiphanies, and failures. The notebook might be divided into sections such as notes, diagrams, models, collected data, assessment questions, and musings. On the left side of the notebook, students make representations of personal connections made during their investigations (Young, 2003).

Figure 1.5 shows various ways a student can represent their knowledge of a particular topic in their interactive notebook (Waldman & Crippen, 2009).

It is usually desirable to allow students to depict their understanding of the phenomenon independently so that misconceptions can be addressed right away. This can be seen in the student work example provided in Fig. 1.6. If the teacher were to only assess the student's knowledge of the phenomenon through an either/or question, then the teacher would have been misled as to what the student actually understood.

By providing space for the student to depict their knowledge through a diagram (see Fig. 1.7), the teacher is able to pinpoint misunderstandings. In this case, the student identified the correct pattern in which temperature and day length varied

Left Side	Right Side
• Drawing, photo, or illustration of new concept or idea.	• Notes taken during class discussion, lecture, or investigation.
• Personal reflections about information.	• Procedure or materials for investigation.
• Predictions, contradictions, or quotations.	• Data collected during collaborative activities.
• Metaphors, analogies, acronyms, poems, songs, or cartoons that represent the new information.	• Ideas generated from activity.
• Connections between the information, and the student's life.	• Diagrams, models, or any illustration that depicts the students understanding of the phenomenon during that instructional sequence phase.
• Summary of activities.	

Fig. 1.5 Ways a student can represent their knowledge in an interactive notebook

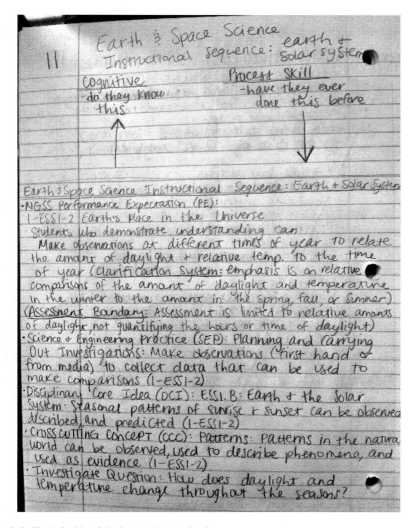

Fig. 1.6 The left side of the interactive notebook

from season to season but demonstrated a poor understanding of the real causes of the phenomenon.

Through this example, the student was offered a space to depict their knowledge and the teacher given a chance to have a deeper insight into real student understanding. In this case, the diagram reveals a misconception that the earth actually moved closer to the sun in summer during its orbit. Through the interactive notebook, the processes leading to knowledge are made more transparent, thereby reducing the likelihood of misconceptions and erroneous conjecture.

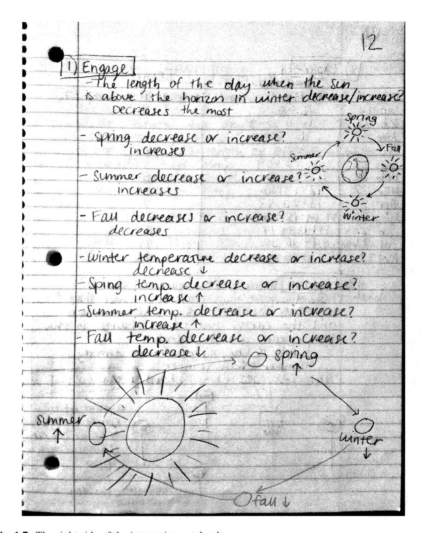

Fig. 1.7 The right side of the interactive notebook

Photo Narrative

Two challenging tasks of the scientist/engineer have always been:

1. Precisely recording documentation
2. Finding the right words to accurately communicate the intricate processes and complex phenomena of scientific investigations

The photo narrative is an easy, contemporary way to help with both challenges. Photo narratives are a way for students to "tell the story" of an experiment and to

1 Inquiry, Investigative Processes, Art, and Writing in STEAM

document the results from following a set of procedures. Dolberry (2010) notes that storytelling is an enjoyable activity that also helps student remember important details that "strengthen their scientific knowledge" (p. 175). The technique is as suitable for students in elementary school as for university students in doctoral degree programs.

Indeed, the photo narrative provides instantaneous documentation in the form of an image for each step of an investigation. Rather than a student trying to recall every detail of long, complicated processes from memory, the photo narrative lets students capture and document each moment as it unfolds. In this manner, the photo narrative segments any experiment or investigation into a series of definable steps. At the end of an experiment, printing out images in sequence also offers the opportunity for a holistic view of the entire project, from start to finish. This view also makes possible the isolation and analysis of specific procedures and effects.

Students must have access to a digital camera or a device that takes photos, such as a smartphone. As an introduction to the assignment, a teacher might ask the class "Why is a picture worth 1,000 words?" and discuss the value of using images to supplement textual description.

Inevitably, students will claim that they "already know" how to shoot a photo, but a short, 5-minute introduction to focus, lighting, and especially distance from the subject is recommended. Students should be encouraged to shoot close-ups (though they will initially tend to shoot everything from too far of a distance). Before students begin work on their narratives, ask them to shoot sample close-up images. A teacher should only approve shots that are in focus and sufficiently close to the subject. Do not accept poor photographs.

The Internet is chock full of "how-to" videos that use images and words to demonstrate how to do something. Examining an effective "how-to" video and discussing traits that make a "how-to" video interesting and useful would help demonstrate to students the necessity of close-ups and sufficiently detailed text. Viewing a woefully ineffective video with shoddy images and vague text might also help students understand the need for precision.

Once students have taken photos of a particular step in the process, a teacher might need to first demonstrate how to write words to accompany photos. Words should not only explain the image but also describe the process. Text should also include observations, commentary, and speculations, which may or may not wind up in the final copy.

The following excerpt of a photo narrative is the work of a high school student in ninth grade, explaining how to solve a Rubik's Cube. The student created 29 photos with textual explanations; 10 are shown in Fig. 1.8. Note the details in the text that extend the information conveyed by only the image.

Photo narratives would be appropriate for most scientific investigations and engineering design processes.

After students create a photo narrative, a capstone assignment is to require them to transform their images and text into an oral presentation. A second possible capstone assignment is to have another student or group of students attempt to replicate the study based upon the descriptions of the photo narrative. The effective photo

	Title page: For this narrative, we will be using the green face as a starting point in solving by forming the green cross. We will learn more about this step later. All algorithms are performed with the cube's front color facing you in the corresponding situation. The color of each face on the cube is determined by the color of the centerpiece.
	Photo 1: The Back Wedge, or B. Solving a Rubic's Cube is an intuitive and algorithmic process. One must use certain algorithms, or fixed set of movements to solve a cube. These algorithms are notated by the wedges of the cube. For example, B is the Back Wedge. Other possible wedges: Front or F, Left Wedge or L, Right Wedge or R, Up Wedge or U.
	Photo 2: The Down Wedge, or D. Movements are indicated by algorithms like this: RD'RD. The letter signifies which wedge to turn. The apostrophe means the wedge is turned counterclockwise. A letter by itself means the wedge is turned clockwise. For example the algorithm R'D'RD would involve turning the right face counterclockwise, the down face counter clockwise, the right face clockwise and the down face clockwise.
	Photo 7: The Scrambled Cube. The first objective in solving the Rubik's Cube is to solve the green face. We will do this by inserting edge pieces (pieces with two colors on them as opposed to corner pieces which have three) into their corresponding positions and lining up the color of each center with each piece. Okay, first step, identify edge pieces with green on them.

Fig. 1.8 Photo narrative on how to solve a Rubik's Cube (ninth grade) (Baines & Kunkel, 2016)

1 Inquiry, Investigative Processes, Art, and Writing in STEAM   15

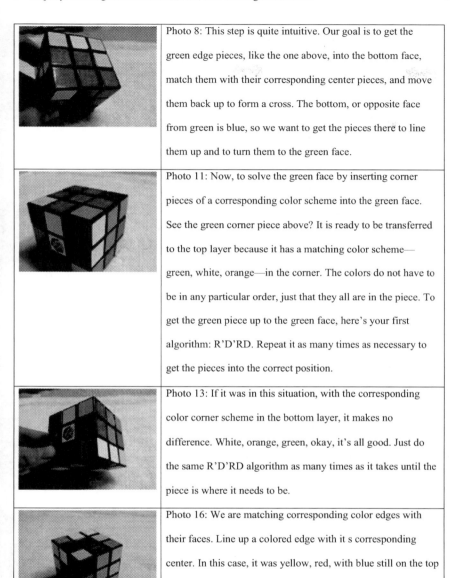

	Photo 8: This step is quite intuitive. Our goal is to get the green edge pieces, like the one above, into the bottom face, match them with their corresponding center pieces, and move them back up to form a cross. The bottom, or opposite face from green is blue, so we want to get the pieces there to line them up and to turn them to the green face.
	Photo 11: Now, to solve the green face by inserting corner pieces of a corresponding color scheme into the green face. See the green corner piece above? It is ready to be transferred to the top layer because it has a matching color scheme—green, white, orange—in the corner. The colors do not have to be in any particular order, just that they all are in the piece. To get the green piece up to the green face, here's your first algorithm: R'D'RD. Repeat it as many times as necessary to get the pieces into the correct position.
	Photo 13: If it was in this situation, with the corresponding color corner scheme in the bottom layer, it makes no difference. White, orange, green, okay, it's all good. Just do the same R'D'RD algorithm as many times as it takes until the piece is where it needs to be.
	Photo 16: We are matching corresponding color edges with their faces. Line up a colored edge with it s corresponding center. In this case, it was yellow, red, with blue still on the top face. I lined up the yellow, red edge with the yellow center because the yellow part of the edge lines up with its

Fig. 1.8 (continued)

	corresponding color. So, we have it matched, now we must get the red color on top of the yellow color on the edge to the left, to match it up and solve the gap. Holding the edge whose color is lined up with the center forward (example: yellow with yellow), do this algorithm: U'L'ULUFU'F', and that should solve it.
	Photo 27: Finally!!! The last step. And easy. It's just like putting in each corner of the cross in the beginning. We're just correctly putting each one corner into a correct orientation, solving the cube. Do R'D'RD on each piece you want until it reaches the top layer, then rotate the U face (Up) until another piece is in position to be put up. Do the R'D'RD algorithm and repeat this process until the cube is almost solved. (When doing this step, it appears you have messed up the cube, but if you stick to it, it will turn out fine).
	Photo 29: Congratulations. You have mastered the Rubik's Cube.

Fig. 1.8 (continued)

narrative would be one that offers clear documentation and leads to similar results. The ineffective photo narrative would be one that leaves out information or gives wrong directions and leads to skewed, erratic, or erroneous results.

Conclusions

Integrating images and the arts into inquiry-based approaches to science and engineering can significantly enhance the quality of student learning. Yet, both of the examples presented in this chapter—interactive notebooks and photo narratives—also require students to engage in a great deal of writing. Indeed, both interactive

1 Inquiry, Investigative Processes, Art, and Writing in STEAM

notebooks and photo narratives require students to use writing as a way of capturing real-time experience. In a review of the effect of writing on learning in science, Reynolds, Thaiss, Katkin, and Thompson (2012) found that writing was an "effective tool in student learning and engagement" (p. 17). By implementing a writing-intensive approach in studying microbes, Burleson and Martinez-Vaz (2011) found that "ninety-one percent of students demonstrated increased knowledge of microbial concepts and methods" (p. 2).

If schools could foment a change in attitudes toward science, technology, art, mathematics, and technology education, they could "become the social institution through which leadership is provided and action is initiated to reconstruct society" (Schiro, 2013, p. 167). These days, STEM education often gets disparaged as too dull, too abstract, and too irrelevant, especially from the perspectives of students (Potvin & Hasni, 2014). Integrating the arts, requiring students to think and write, and stimulating students' creative impulses can promptly dispel any notions of ennui and help immeasurably in transforming STEM into a thoroughly contemporary STEAM.

References

Aulls, M. W., & Shore, B. M. (2008). *Inquiry in education*. New York: Lawrence Erlbaum Associates.

Baines, L., & Kunkel, A. (2016). *Going bohemian*. Seattle, Wa: Amazon Digital Services.

Bruner, J. S. (1961). The act of discovery. *Harvard Educational Review, 31*(1), 21–32.

Burleson, K. M., & Martinez-Vaz, B. M. (2011). Microbes in Mascara: Hypothesis-driven research in a nonmajor biology lab. *Journal of Microbiology & Biology Education, 12*(2), 166–175.

Bybee, R. W. (1982). Historical research in science education. *Journal of Research in Science Teaching, 19*(1), 1–13.

Capra, F. (2008). *The science of Leonardo*. New York: Doubleday.

Chesbro, R. (2006). Using interactive science notebooks for inquiry-based science. *Science Scope, 29*(7), 30–34.

Dewey, J. (1933). *How we think*. Boston: DC Heath.

Dewey, J. (1934). *Art as experience*. New York: The Berkley Publishing Group.

Dewey, J. (1938). *Experience and education*. New York: Collier.

Dolberry, A. A. (2010). The sci-fi microbe: Reinforcing understanding of microbial structures and their significance through a creative writing exercise. *Journal of Microbiology & Biology Education, 11*(2), 175–176.

Greene, M. (1995). *Releasing the imagination*. San Francisco: Jossey-Bass.

Hammerman, E. (2006). *8 essentials of inquiry-based science*. Thousand Oaks, CA: Sage Publications.

Hawkins, J., & Pea, R. D. (1987). Tools for bridging the cultures of everyday and scientific thinking. *Journal of Research in Science Teaching, 24*(4), 291–307.

Jaladanki, V., & Bhattacharya, K. (2015). Arts-based approach to physics instruction. *Creative Approaches to Research, 8*(2), 32–45.

Karplus, R., & Their, H. D. (1967). *A new look at elementary school science*. Chicago: Rand McNally.

Maeda, J. (2013). Artists and scientists: More alike than different. *Scientific American*. https://blogs.scientificamerican.com/guest-blog/artists-and-scientists-more-alike-than-different/.

Massiala, B. G. (1969). Inquiry. *Today's Education, 58*, 40–42.

National Research Council (NRC). (2012). *A framework for k-12 science education: Practices, crosscutting concepts, and core ideas* (Committee on a Conceptual Framework for New K-12 Science Education Standards. Board on Science Education, Division of Behavorial and Social Sciences and Education). Washington, DC: The National Academies Press.

Piaget, J. (1960). *The psychology of intelligence*. Totowa, NJ: Littlefield, Adams & Co.

Potvin, P., & Hasni, A. (2014). Interest, motivation and attitude towards science and technology at K-12 levels: A systematic review of 12 years of educational research. *Studies in Science Education, 50*(1), 85–129 http://www.tandfonline.com/doi/abs/10.1080/03057267.2014.881626

Reynolds, J. A., Thaiss, C., Katkin, W., & Thompson Jr., R. J. (2012). Writing-to-learn in undergraduate science education: A community-based, conceptually driven approach. *CBELife Sciences Education, 11*, 17–25.

Robelen, E. (2011, December 7). STEAM: Experts make case for adding arts to STEM. *Education Week*. http://www.edweek.org/ew/articles/2011/12/01/13steam_ep.h31.html

Schiro, M. S. (2013). *Curriculum theory: Conflicting visions and enduring concerns*. Thousand Oaks, CA: Sage Publications.

Schon, D. A. (1992). The theory of inquiry: Dewey's legacy to education. *Curriculum Inquiry, 22*(2), 119–139.

Waldman, C., & Crippen, K. J. (2009). Integrating interactive notebooks. *The Science Teacher, 76*, 55–59.

Wheatley, G. (1998). *Problem-centered learning*. Tallahassee, FL: Florida State University.

Young, J. (2003). Science interactive notebooks in the classroom. *Science Scope, 26*(4), 44–47.

Chapter 2
Structuring STEAM Inquiries: Lessons Learned from Practice

Sarah B. Bush and Kristin L. Cook

Structuring STEAM Inquiries: Lessons Learned from Practice

Elementary teachers are usually responsible for teaching all content areas and could therefore benefit from professional development on cross-curricular planning and instruction. Research on integration suggests that "a large number of K-12 studies sustain the notion that integration helps students learn, motivates students, and helps them build problem-solving skills" (Czerniak, 2007, p. 545). Hurley's (2001) meta-analysis of integrated instruction recognizes appeal for integrated method courses offered by teacher preparation programs, stating the merit of integrated and thematic approaches to curriculum. She found that most empirical research supports integrated instruction, suggesting as well that integration fosters learning, motivation, and problem-solving skills. Park Rogers and Abell (2008) also provide a concise review of benefits of integrated instruction, including maximizing instructional time, reinforcing concepts, learning big ideas, and fostering cross-curricular connections. They note concerns with integration, however, such as an unequal focusing on one discipline more than others or lack of powerful instruction in any one topic because all boundaries are blurred. Among the various content areas, integration of science and mathematics has received much attention (Czerniak, 2007). In their review of science and mathematics integration literature, Pang and Good (2000) found that mathematics is often integrated into science instruction as an adjunct component to science content. Moreover, while the rationale for science and mathematics integration is clear (e.g., they require similar inquiry and

---

S. B. Bush (✉)
University of Central Florida, Orlando, FL, USA
e-mail: sarah.bush@ucf.edu

K. L. Cook
Bellarmine University, Louisville, KY, USA

© Springer Nature Switzerland AG 2019
M. S. Khine, S. Areepattamannil (eds.), *STEAM Education*,
https://doi.org/10.1007/978-3-030-04003-1_2

problem-solving skills), the actual implementation of integrated instruction in the classroom is rare (Watanabe & Huntley, 1998).

Through integrated instruction, elementary teachers can simultaneously address standards across content areas. Unfortunately, most elementary school curricula are currently disjointed and isolated, with certain time frames dedicated to each subject and little integration (Watanabe & Huntley, 1998). Teachers can be taught to integrate their curricula and be given opportunities to reflect on how to do so through ongoing professional development. Park Rogers and Abell (2008) suggested professional development ought to include a focus on process skills, the use of national and state standards to drive the planning of thematic units, and the use of strong and meaningful themes. Therefore, in our work, content standards are intentionally integrated with value importantly given to meeting the goals of each of the five content areas in STEAM: science, technology, engineering, the arts, and mathematics. Furthermore, our professional development efforts encouraged integration that was driven by teachers' objectives (viz., state content standards).

Our work with in-service elementary teachers is grounded in the research base which supports the effectiveness of integrated STEM instruction (such as in Becker & Park, 2011; Hom, 2014). Integrated STEM instruction has shown positive effects on student achievement as well (Becker & Park, 2011). Read (2013) argues that the majority of K-8 teachers are underprepared in their mathematics and science content and that they need more training in both content and practices. A report from the Early Childhood STEM Working Group (2017) recommends a revamping of in-service STEM-related training for teachers of young children and highlights the importance of making high-quality STEM resources for young children available to practitioners. To prepare students for the twenty-first-century demands, a truly integrated approach to teaching the STEM subjects is needed as we prepare teachers to teach integrated STEM so that they, too, can understand the connections among the STEM disciplines (Ostler, 2012). Our program works to help teachers learn how to integrate mathematics and science concepts together into cohesive STEM curricula using high-quality resources, a skill which Lewis, Alacaci, O'Brien, and Jiang (2002) have documented is often challenging for teachers.

In addition to developing in-service elementary teachers' abilities to effectively integrate the STEM subjects meaningfully, we are exploring how the integration of the arts can be used as a "hook" or a way to engage more diverse learners – often those that have not typically been drawn to the individual STEM subjects. A new, but rather limited, body of literature has emerged which supports the benefit of arts integration into STEM as a way to engage more types of learners (Ahn & Kwon, 2013; Bequette & Bequette, 2012; Wynn & Harris, 2012). The addition of the "A" taking it from STEM to STEAM recognizes the role of aesthetics, beauty, and emotion to arriving at a solution to a problem (Bailey, 2016). Incorporating the arts adds a needed affective component to complex STEM concepts and problems, often making it more accessible (Peppler, 2013; Smith and Paré, 2016) and more engaging. A recent study conducted by Herro and Quigley (2016) studied the effectiveness of a multiyear STEAM professional development program and found that through this program teachers increased their understanding of STEAM. Although the research on the effectiveness of STEAM education is limited, more districts and schools are engaging in teaching integrated STEAM each year (Delaney, 2014). Our work aims to add to the limited literature base on integrated STEAM education.

People think about STEM or STEAM in many different ways. Bybee (2010) argues that STEM curricula should be set in a context and aligned to real-life issues that can be addressed through each of the four STEM areas. Read (2013) describes STEM as "education in math or science, using engineering design approaches and technology tools, delivered through a combination of hands-on, student-centered, inquiry-based projects and direction instruction" (slide 3). In our work with STEAM education, we agree with Herro and Quigley (2016) that STEAM incorporates the idea of transdisciplinary learning which is the idea that students learn through a true blending of the disciplines and that they are solving problems set in a real context (as in Klein, 2014). In transdisciplinary teaching, students become so engaged in solving the problem that they are excited to draw on prior knowledge and learn new concepts from the different STEAM disciplines in order to reach a solution. We work with teachers to develop authentic curricula where students are engaged in working together to solve a real problem, and in order to solve the problem, they must synthesize their knowledge of the STEAM disciplines to reach a viable solution. Our project also draws on problem-based inquiry which has been shown to improve urban and minority students' achievement and engagement in learning in math and science (Buck, Cook, Quigley, Eastwood, & Lucas, 2009).

Full STEAM Ahead: Project Overview

Project Background

In spring 2015, approximately six months prior to the start of our professional development program, we began collaborating with the elementary science and elementary mathematics content specialists at our partner school district. We learned about district priorities and current changes and practices in mathematics and science assessments including that our state was moving to a completely revamped state standardized science assessment. During this time we identified five elementary schools in the district that showed an interest in launching STEAM efforts but needed direction and help to build infrastructure and sustainability. We also set the following project goals: *(1) increase students' science and mathematics achievement, (2) increase teachers' and instructional coaches' science and mathematics pedagogical content knowledge, and (3) build a community of educators dedicated to STEAM teaching and learning.*

Additionally, we established partnerships with a mathematician and biologist in the arts and sciences college at our university. We also formed partnerships with three informal learning partners in our city including a state science center, a center for performing arts, and a art musuem involved in STEAM. Our informal learning partners served as experts in innovation and provided some of the professional development as well as function as a resource for participants and as part of our community of STEAM educators. We also collaborated with a consultant who had experience in creating STEM centers throughout the United States and internationally. Finally, we invited two expert K-5 STEM/STEAM lab teachers to serve on our leadership team providing a critical and current practitioner lens to the leadership group. The lead facilitators, who were also the PIs on the project, were a mathematics teacher educator and science teacher educator who aim to explore the effectiveness of truly integrated STEAM instruction. Finally, we had the expert guidance from a team of external evaluators.

By the time the project, funded by a Mathematics Science Partnership (MSP) grant, began in fall 2016, we had enlisted five schools from our large urban partner district from the Midwest. Within those 5 schools, 25 classroom teachers (all grades 3–5 except 1 special education teacher and 1 second-grade teacher), 5 STEAM instructional coaches (1 from each school), and at least 1 building administrator for each school were signed on as the project participants. The STEAM instructional coach from each school served as the school leader and was responsible for leading school-level professional learning communities (PLCs) and organizing the classroom implementation portion of the professional development. Participants had a wide range of teaching experience, from 2 to more than 20 years, and educational attainment ranging from an initial teaching certification to multiple advanced certifications. Some participants had a variety of experience in terms of different schools, districts, and grade levels they had taught, while others had spent their entire teaching career (thus far) in one classroom placement.

Our professional development schedule was developed with research-based qualities of effective professional development in mind. For example, while we knew it would be complex and at times challenging, cyclically connecting professional development to classroom implementation (Desimone, 2009; Loucks-Horsley, Stiles, Mundry, Love, & Hewson, 2010; McAleer, 2008; Sztajn, 2011) was essential to pushing integrated STEAM forward in our participating schools. In order to create iterative cycles where teachers participated in whole-group professional development and then went back to their classrooms to implement new strategies, our professional development program took place from October through April each year. We met as a whole group approximately two times each month with ongoing classroom implementation.

Additionally, each school-level PLC met in their building. In addition to the structure of the schedule, we also knew it was important to situate the learning during the professional development in the context of the participants' classroom and school environment (Putnam & Borko, 2000). When teacher learning becomes "situated," the teacher can begin to alter their own teaching practices in alignment with the professional development which can result in changes and growth in their skills and knowledge of the practices of teaching (as in Borko, 2004; Greeno, Collins, & Resnick, 1996; Lave & Wenger, 1991). This change and growth can take on an iterative cycle of its own – as teaching knowledge and skills improve, teachers have new knowledge and skills to offer during the professional development sessions, which in turn improves the quality of the professional development for the group (and the individual), which continues to improve the knowledge and skills for teaching, and the cycle continues. In this type of professional development environment, teachers engage with professional development leadership to help co-construct the learning experiences, which is different from traditional professional development where teachers are only seen as participants (or recipients) (Timperley, 2011).

A Focus on Content Standards and Practices

At the foundation of our work with teachers was a critical focus on mathematics and science content and practices as outlined by the Common Core State Standards for Mathematics (CCSSM; CCSSO, 2010) and the Next Generation Science Standards (NGSS; NGSS Lead States, 2013). We focused specifically on standards for grades 3 through 5 and worked to target content areas identified as areas of unfinished learning in mathematics achievement for students in our participating schools. This helped us plan our professional development to best meet the needs of the participating schools. All of our professional development was conducted through the primary focus of the CCSSM and NGSS content and practice standards with meaningful connections made to art and technology standards.

We also guided our project participants in being explicit and focused on the alignment of the CCSSM and NGSS content and practices as they planned the STEAM inquiries they would implement into their classes. To accomplish this, we required participants to document the specific CCSSM and NGSS content and practices, as well as art and technology standards that were being addressed in a planning template we created. Participants used curriculum pacing guides from their school district to determine CCSSM and NGSS standards of focus. At first, many participants were aligning their STEAM inquiries to every standard that might have a connection, which tended to have a "mile wide and inch deep" effect. We guided teachers to only align those standards that were of primary focus and which were assessed through their planned STEAM inquiries.

How well the CCSSM and NGSS content and practices as well as art and technology standards were being addressed during classroom implementation of STEAM inquiries was a focus of school visits that occurred during year 2 of the project. During this time, both project leaders and participants observed STEAM inquiries of project participants in their classrooms and used an observation tool we created to document the strength of the alignment of the inquiries to appropriate CCSSM, NGSS, art, and technology standards.

Housing STEAM inquiries within problem-based scenarios enabled teachers to encourage collaborative problem-solving that required the use of and knowledge of various content areas as well as set an authentic context within which to explore. Problem-based learning (PBL) is a process that promotes learning through working together to solve a real-life problem. Students practice science in the classroom the way that scientists and engineers do working in collaborative groups to iteratively solve problems and explore challenges (Savery, 2006). As such, the benefits of using PBL include but are not limited to increased content knowledge, higher-order thinking, self-directed learning, and twenty-first-century skills such as collaboration, creativity, and critical thinking (Hmelo-Silver, 2004). For our purposes, PBL offered an entry point for teachers embarking on designing meaningful and complex STEAM inquiries.

Creation of STEAM Inquiries

We used an adapted version of the problem-solving cycle (PSC; Borko, Jacobs, Koellner, & Swackhamer, 2015) to plan STEAM inquiries in whole-group professional development sessions, implement them in the classroom, and then reflect on the instruction during the following whole-group professional development. We worked with participants to employ Bryk, Gomez, and Grunow (2010) "Plan, Do, Study, Act" cycles so that participants essentially engage in and conduct action research. At the beginning of each cycle, during a whole-group professional development session, school- and grade-level PLCs determine a "change idea" they want to test during implementation of their STEAM inquiry. They work as a team to develop their STEAM inquiry and complete a planning document (mentioned above) that includes key alignment to CCSSM, NGSS, art, and technology standards. During this whole-group professional development session, groups meet with leadership team members that specialize in mathematics, science, arts, and technology for guidance. They plan as a group and share plans with the leadership team. They also meet with fellow project participants that will be observing their classroom in order to debrief on the purpose and content of the planned inquiry. Often teachers leave the whole-group session with a fairly well-formed plan, and they finalize the details and gather materials once back in their school building prior to implementation.

During implementation, participants are observed by selected leadership team members and fellow participants. All observers complete a STEAM observation feedback form we created for the project which is used to guide a discussion of the lesson's effectiveness at the following whole-group session. Teachers collect artifacts from their lesson, such as student work and pictures. They bring their planning documents, STEAM observation feedback forms, and lesson artifacts as evidence to the following whole-group session.

At the whole-group session following classroom implementation, the session is dedicated to structured debriefs of each lesson observation. Observer/observe teams meet for structured time blocks to discuss both student thinking and the teacher instructional strategies (as in Borko et al., 2015) through the lens of the STEAM observation feedback forms, planning documents, and lesson artifacts. These post-observation reflection discussions are rich with detail on CCSSM and NGSS content and practices as well as pedagogical strategies, student engagement, relevancy, questioning, the level of true integration, and more.

Three Classroom-Tested Inquiries

In this chapter, we provide readers with three classroom-tested inquires that can serve as helpful exemplars of meaningful integration of STEAM using authentic and engaging contexts. These three examples include the following STEAM inquiries: designing a prosthetic arm for a kindergartener, a paleontology investigation, and a closer look at the arts within roller-coaster engineering. In each case we describe the context and the inquiry itself, and then we focus on the alignment to key standards, with a critical focus on the CCSSM and NGSS content and practices. The first inquiry, designing a prosthetic arm for a kindergartener, highlights the use of technology as a strategic tool to increase the effectiveness of instruction. The second inquiry, a paleontology investigation, highlights how scientists use science and mathematics to recreate extinct animals. The third investigation focuses on the benefits of the arts integration into roller-coaster engineering.

Three Classroom-Tested Inquires

Designing a Prosthetic Arm for a Kindergartener

The driving prompt for this fourth-grade STEAM lesson was inspired by a local kindergartener who was missing a portion of her right arm and thus had difficulty logging onto school computers. By synthesizing the various content areas of STEAM, students were asked to design and build a prosthetic arm that would enable the student to simultaneously press the Control-Alt-Delete buttons on the keyboard. Motivated by this authentic and meaningful context, students began by undergoing a series of tasks that would help them empathize with the student for whom they would be designing. They explored the American with Disabilities Act and conducted an inventory of tasks they could or could not accomplish around the school (i.e., using the restroom, washing hands in the sink, opening doors, carrying books from locker to classroom, etc.) and submitted their conclusions to the principal about the accessibility in their school. Through this guided exploration, students became aware of how difficult simple tasks are with only one arm and understood the seriousness of the project on which they were about to embark. Purpose-driven and highly engaged in their task, students conducted research about prosthetics and anatomy/skeletal system. They also investigated ways in which other animals use body parts as hands, such as how opossums use their tails to grasp onto trees and how geckos have sticky substances on their fingers to adhere to surfaces. These explorations of nature helped students generate ideas for their prototype, which they drew as a schematic in their science notebooks. Students were excited to begin building their prosthetic using the ideas they generated from their research, though

Fig. 2.1 Students considering scale in their designs

the teacher emphasized how the design process is a series of planning/replanning/pilot tests before the actual fabrication of their prototype on the classroom 3D printer. First, students were asked to create a blueprint on Tinkercad, an online site for creating digital designs that can ultimately be printed three dimensionally. In doing so, students considered measurement and scale of their designs (see Fig. 2.1). Next, students continued to work in teams to build a physical prototype of their design using simple hardware the teacher provided for construction. Through presentations, groups had to sell their prototype as the "design of the day" that would be the final prosthetic printed when the groups merged as one design company. Stakeholders such as the principal, the district technology integration specialist, other teachers, and classmates served as the audience for these presentations. The final prosthetic was then printed and given to the family of the kindergartener. This meaningful learning experience engaged students in solving a problem that was important to them, and their excitement about being able to help a local family drove active participation in the long-term project.

Completing this task necessitated that students explore and deepen their understanding of the various content areas embedded in STEAM. With regard to science content and practices, students focused on structure and function (plants and animals have both internal and external structures that serve various functions in growth, survival, behavior, and reproduction 4-LS-1). Students researched the skeletal and muscular system and understood that missing appendages resulted in the need for alternative ways of completing tasks important for survival. They designed their prosthetic in a way that considered how animal species interact with environment. Building models and construction of explanations about their designs allowed students to immerse themselves in these important science and engineering practices. At the heart of this inquiry, students defined a simple design problem reflecting a need or want that includes criteria for success and constraints (3-5-ETS1-1). Being successful in this project (i.e., building a functional prosthetic that completed the task it was charged to perform as well as fit and attached to the kindergarteners' arm) required students to consider length and angle measurement (4.MD.6), measurement conversions between customary and metric and within metric (4.MD.1), and using the four operations with decimals (4.MD.2) to develop a budget for team supply list. With regard to the mathematics practices, students had to make sense of

2 Structuring STEAM Inquiries: Lessons Learned from Practice

problems and persevere in solving them (SMP 1), construct viable arguments and critique the reasoning of others (SMP 2), and attend to precision (SMP 6). The technology in this lesson (i.e., 3D printer and use of Tinkercad design software) was used authentically in the design process to create a blueprint and allow for fabrication of the design. Rather than using technology for the sake of technology, the seamless use of technology facilitated the task rather than took over the task. The 3D printing actually took place at night as the product, and not the printing itself, was of interest. The art integration centered on drawing, proportions, and scaling (also a mathematics connection) of the design. Aesthetics of designing a prosthetic (structure, design, color, overall look) were important, as the class was hopeful the student for whom they were designing would want to wear their product. As it turned out, the student loved their design though she did request a different color (pink instead of black)! More details related to this inquiry can be found in Cook, Bush, and Cox (2015) and Bush, Cox, and Cook (2016).

A Paleontology Investigation

This fourth-grade inquiry used embedded STEAM content to solve a paleontology-related dilemma. To encourage students' questioning, critical thinking, and problem-solving abilities, students were given the central problem: "How can we determine what an entire dinosaur looked like when all we have is its skull?" In their exploration of the problem statement, students examined a skull fossil to determine what the rest of the dinosaur must have looked like. From that, they deduced how the dinosaur must have moved, eaten food, and behaved in its habitat. In this weeklong unit, students used biological, earth science, and mathematics content to artistically design the hypothetical form and function of the dinosaur. As part of the problem statement, students were also asked to make a group presentation to the "Academy of Paleontology" about their findings, prompting them to engage in twenty-first-century learning skills such as collaborative problem-solving and creative critical thinking which were honed during the research and presentation portion of this STEAM inquiry. To begin, student groups (grouped into research teams of three, with each student given a specific role as either (1) *facilitator* who moderated team discussion, kept the group on task, and distributed work, (2) *reporter* who served as the group spokesperson to the class or instructor and summarized the group's activities and/or conclusions, or (3) *recorder* who logged group discussion in a science notebook and recorded data, claims, and evidence) were shown a $\frac{1}{4}$ scale model of a dinosaur skull. Other materials students were initially introduced to included photos of present-day animals with heads similar to the dinosaurs' (i.e., forward-facing eye sockets, canine-like teeth, snout-shaped nose) illustrations of other various dinosaur skeletons, information about bipedal versus quadruped dinosaurs, and detailed photos of the skull from multiple viewpoints with some background detail for scale reference (see Fig. 2.2). In solving the problem, students were asked to

Fig. 2.2 Student considering scale as they work with skull photo

include in their presentations (1) a description of the external features of the skulls, (2) evidence to identify which type of dinosaur it might be, and (3) conclusions about the life of the dinosaur with regard to its survival, growth, behavior, and reproduction. Students conducted research to construct their argument and worked as a team to develop their presentation to the "Academy of Paleontology."

Science, technology, engineering, art, and mathematics were not discrete subjects taught in this lesson but rather tools to be used in the context of the problem to be solved. Specifically, the focus of this lesson was on constructing an argument that dinosaurs have internal and external structures that function to support their survival, growth, behavior, and reproduction (4-LS1-1). By examining the skull fossil, students observed the position of its eye sockets, which indicated the dinosaur was a predator. By using their provided resources, students were able to compare the position of eye sockets in a variety of animals which showed a pattern of predators having forward facing eyes and prey having side-facing eyes. Students also noted the teeth on the skull and compared the teeth to other dinosaurs in the pictures provided. They discussed how the canines were comparatively small and the presence of grinding molars to conclude the dinosaur was an herbivore. By collecting data from the skull and comparing and contrasting the features to extant animals as well as dinosaur fossils, students were able to construct their argument that the dinosaurs' features unveiled how it functioned in its environment. The multiple iterations of the engineering of the dinosaur skeleton addressed the engineering design standard, Generate and compare multiple possible solutions to a problem based on how well each is likely to meet the criteria and constraints of the problem (3-5-ETS1-2).

Beyond the science, students also used key mathematics. With regard to mathematics, students engaged in CCSSM mathematical practices Reason Abstractly and Quantitatively (SMP 2) and Use Appropriate Tools Strategically (SMP 5) and applied key content including measurement, measurement conversation (4.MD.1), and working with scale (5.NF.5). Students were told the skull was a $\frac{1}{4}$ scale model of a dinosaur, and they were tasked with figuring out the length of the whole dinosaur. Students determined the skull was approximately 15 in. long after they measured the height of the 3D-printed skull and multiplied that length by 4. From there, the teacher guided them in measuring the head to body ratio of other dinosaur fossil photos, so students could discover that the head to body ratio tended to be between 10 and 12 times the head (i.e., the body was somewhere between 10 and 12 times the length of the head). Students had to then convert inches to feet to determine the dinosaur must be in the range of 12–15 ft. By using appropriate measuring tools as well as reasoning quantitatively, students were able to solve the problem presented to them. Finally, the arts were another area of emphasis in this STEAM unit. Art integration included many core aspects of arts instruction including creating, performing, presenting, responding to, and connecting different arts forms including dance, media arts, music, drama, and the visual arts. Various improvisational exercises and dance moves, such as working together in small groups to form the shape of a dinosaur and performing a dinosaur "ballet" that included a prehistoric soundscape created by found objects, helped students consider how the dinosaur "acted" or functioned in its environment. The sketches of dinosaur designs in students' science notebooks also included expressions of form and function based on evidence as well as conjectures about what the skin color and texture of the dinosaur might have been like. In this unit, integrating STEAM content helped students construct their argument and consider evidence as a paleontologist does. By taking on the role of a paleontologist with limited data, students came to understand how science is connected to other disciplines and how knowledge from other disciplines is essential to solving complex and real life problems. All specific activities and assessments for this inquiry can be found in Hunter, Cox, Bush, Cook and Jamner (2017) and Cox, Hunter, Cook, and Bush (in press).

A Closer Look at the Arts Within Roller-Coaster Engineering

This fourth-grade STEAM inquiry builds on existing roller-coaster lessons by emphasizing the arts throughout the design process, as the focus is for students to use their imaginations and engineering skills to create their own designs while developing an understanding of energy. Oftentimes, the arts within the context of STEAM get reduced to craft projects with an aesthetic focus. The art integration within this lesson, however, draws upon the design element of roller-coaster engineering through imaginative visualizations, creative story-telling, and an emphasis on careers that combine elements of visual arts with STEM subjects. To begin, students are asked to visualize and imagine what a roller coaster feels like through mental imagery. Students articulate the feelings associated with riding a roller coaster, such as fear and excitement. This visceral exercise promoted engagement in the learning experience as students began to think about the thrill of riding a roller coaster and the points at which they found the most enjoyment during their ride. After perusing several images of different types of roller coasters, all which are designed with a central theme in mind (i.e., Space Mountain, dragon-shaped coasters, and coasters that simulate Indiana Jones' adventures), students are prompted to consider how the storytelling element is present in the roller coaster's design. Throughout the STEAM lesson, students are learning about energy transfers and how speed is connected to energy changes. As they do this, they consider how roller coasters utilize energy changes to maximize the narrative of their story. There are opportunities to discuss many artistic elements of roller-coaster design – from the structure (exterior, interior, and free standing) to trick elements (banked turns, bunny hills, camelbacks, track switches, and vertical drops) to variations (dueling coasters, racing coasters, Mobius loop coasters, and shuttle coasters). This problem-based learning activity is guided by the prompt of students designing a themed roller coaster for a local amusement park that is cost-effective, fun, and safe. Criteria and constraints are given to the students such as a budget, materials, safe and smooth stops, and required number of loops. As ideas about Newton's law of inertia, friction, and velocity are explored, students are asked to include a thematic or storytelling element into their roller-coaster designs. Student teams considered cultural elements, mythical stories, fairy tales, and more to add their own themed environments. In addition to the creative storytelling aspect of the engineered roller coasters, students are also introduced to Walt Disney's "Imagineering," which blends imagination with engineering (see https://disneyimaginations.com/about-imaginations/about-imagineering/). Imagineering allows educators to demonstrate that STEAM-related disciplines can lead to creative careers. By exploring how innovators and makers who work for Disney overcome failure to achieve success, students reflect on their own experiences persevering through challenging tasks. This discussion helped students focus on creative thinking and multiple approaches to design while laying the foundation for overcoming failure, which is a key part of the inquiry. These art-inspired elements show students that a roller coaster can provide a visceral experience and art and storytelling provide an emotional experience. The inclusion of the arts helped students learn that manipulation of the visual aesthetics affect the energy flow of the system, which then determines its functionality – ultimately leading to the development of a more thoughtfully conceptualized and purposeful roller-coaster design.

The art integration in this STEAM inquiry helped underscore and provide experiential meaning to the key mathematics and science content and practices. The primary performance expectation for science was that students use evidence to construct an explanation relating the speed of an object to the energy of that object (4-PS3-1). An online animation allowed students to view the energy exchanges on a smart board and mark where they predicted potential and kinetic energy would be highest. Students discussed Newton's law of inertia that states an object stays in motion or at rest unless acted upon by an outside force and were required to demonstrate their understanding of energy changes throughout their roller-coaster designs. Students considered what factors or materials on their roller coasters were acting as outside forces to slow the marbles (e.g., track, sandpaper, rubber strips, tape, pipe cleaners, and wind resistance). By planning and carrying out fair tests in which variables are controlled and failure points are considered to identify aspects of a model or prototype that can be improved (3-5-ETS1-3), students engaged in the engineering practices of roller-coaster creators. Discussion about what constituted a fair test was an important group consideration. Students had to decide how they would measure the marble's run on the track (mathematics integration), how they would calculate the speed of their track (mathematics integration), and how they would determine the safety of the track consistently so that groups could compare their data (mathematics integration). Students maintaining a budget (4.MD.2) that included a running log of items purchased and costs of materials employed core mathematics principles. Students also were required to calculate overall velocity in the system by dividing the time of their run by the length of the track (7.NS.3). This calculation aided in the students being able to quantify the speed and fun-factor of their tracks and required they use appropriate measuring materials and units in their calculations. The support of online simulations helped students visualize changes in potential and kinetic energy. Thus, the use of technology in this STEAM inquiry was meaningfully integrated to build a more thorough understanding of the abstract physics embedded in the roller coaster's design, enhancing instruction. For more specific details regarding this inquiry, see Cook, Bush, and Cox (2017).

Suggestions for STEAM Implementation

Lessons Learned

During the past several years, we have spent much time reflecting on our work with in-service elementary teachers and their students as we move toward the common goal of implementing STEAM instruction in a way that is truly integrated and trans-disciplinary in nature, as well as authentic and meaningful. Through our work, we have found one of the most powerful aspects of integrated STEAM that separates it from integrated STEM is the incorporation of the empathy (see Bush & Cook, 2019 for more information). When teachers implemented STEAM lessons that sought to problem solve on behalf of others or design with someone else in mind, their sense of purpose and engagement in the lessons increased. As a result, we have shifted to structuring many of our inquiries using the design-thinking framework (Institute of Design at Stanford, 2016), which underscores the intent of the science and engineering practices called for in the NGSS but also begins inquiries with students empathizing with the situation or with others. For example, as noted above in our *Designing a Prosthetic Arm for a Kindergartener* exemplar, students tried to complete many typically daily tasks at their school using only one arm. They quickly found this to be extremely challenging and began to empathize with the kindergartener who faced this reality daily. Building and creating based on the foundation of empathy bring the idea of caring into the classroom and makes for more meaningful STEAM explorations. We advocate that STEAM investigations need not be planted or contrived real-life scenarios – instead they can be actual authentic scenarios facing the communities in which students live.

Tips for Practice

Teaching in a way that meaningfully integrates the STEAM subjects is drastically different from traditional teaching and can be challenging at times. We offer the following tips to serve as a guide for readers ready to embark on transdisciplinary STEAM instruction:

- Children sometimes struggle with the idea of empathy. In our work, we have found the empathy piece of integrated STEAM instruction to be extremely powerful. It is worth the extra time and energy to engage students in this type of thinking because it builds students' motivation and passion toward finding a successful solution to the problem. We suggest using a task as we did with the *Designing a Prosthetic Arm for a Kindergartener* inquiry, videos, research, children's literature, and meaningful classroom discourse as you build the empathy piece in your STEAM inquiries.
- Teaching STEAM in an authentic and transdisciplinary way is vastly open-ended and complex. It is inevitable that students will enter the problem and arrive at a solution using many different paths. This type of instructional environment requires a great amount of flexibility on the part of both the student and the teacher. In this environment, there is no way to anticipate every question a student may ask, every piece of knowledge from each content area that a student may access, or the direction the inquiry may take. Embrace this chaos, and model the aspects of lifelong learning and curiosity with your students.
- Think about the tools and resources your students might find helpful for the STEAM inquiry in which you are about to embark. Try to have as many different resources and materials available to students as possible. You do not need expensive technology or fancy materials to engage students in meaningful STEAM learning. Oftentimes, household materials and resources borrowed from other teachers in your school will suffice. Teachers may also serve as resources. You might find benefit from collaborating with teachers in your school who have expertise to share with students related to your investigation. Outside experts in your community might also wish to get involved!
- Take this teaching transformation 1 day at a time. Start by planning and implementing one integrated STEAM inquiry, which may be a 1-day or a weeklong

inquiry. Take advantage of a current problem that needs to be solved in your community – seize the moment!

- Try to focus on central concepts and practices. STEAM already incorporates many content areas and their accompanying standards. Trying to include too many performance expectations or standards that may be only touched upon can result in unclear directions and expectations for students. Think about what you wish to assess and ensure that tightly aligns to the standards in which you choose to focus your STEAM inquiries.
- Consider classroom norms and expectations. Problem-based STEAM learning is highly collaborative in nature. It also requires perseverance as students embark on challenges for which their proposed solutions often fail. It is important that teachers support their students' work in such an environment by providing ideas for how to overcome failure points and work in a cooperative setting with peers. Not doing so may impede on an otherwise great STEAM inquiry.

Concluding Remarks

Elementary teachers in particular face many pedagogical demands related to the teaching of content and practices of a variety of subject areas. We believe the learning that occurs though a STEAM problem-based context yields many positive benefits. Although there are many ways for schools to structure these experiences, we have seen firsthand the power of having a STEAM lab that deepens and extends the science and mathematics learning in the regular classroom. If schools choose this approach, it is essential the classroom teacher is a part of the STEAM lab experience alongside their students. For example, many classroom teachers co-teach STEAM in a lab with the lab teacher. In this way, classroom teachers can underscore and hone the content related to the STEAM inquiries while the science, mathematics, and engineering practices are extended in the lab. Explorations in problem-based STEAM inquiries take time and require students to synthesize ideas and work collaboratively to solve real-world problems; as such, engaging in these types of learning environments complements the goals for twenty-first-century learning. The power of STEAM teaching and learning derives from the aim to improve life and solve problems through innovation, design, and creative thinking.

References

Ahn, J., & Kwon, N. (2013). An analysis on STEAM education teaching and learning program on technology and engineering. *Journal of the Korean Association for Research in Science Education, 33*(4), 708–717.

Bailey, C. (2016). An artist's argument for STEAM education. *Education Digest, 81*, 21–23.

Becker, K., & Park, K. (2011). Integrative approaches among science, technology, engineering, and mathematics (STEM) subjects on students' learning: A meta-analysis. *Journal of STEM Education: Innovations and Research, 12*(5), 23–37.

Bequette, J., & Bequette, M. (2012). A place for ART and DESIGN education in the STEM conversation. *Art Education, 65*(2), 40–47.

Borko, H. (2004). Professional development and teacher learning: Mapping the terrain. *Educational Researcher, 33*, 3–15.

Borko, H., Jacobs, J., Koellner, K., & Swackhamer, L. E. (2015). *Mathematics professional development: Improving teaching using the problem-solving cycle and leadership preparation models*. New York: Teacher's College Press (copublished with NCTM.

Bryk A. S., Gomez L. M., & Grunow, A. (2010). *Getting ideas into action: Building networked improvement communities in education*. Stanford, CA: Carnegie Foundation for the Advancement of Teaching. Retrieved from http://cdn.carnegiefoundation.org/wp-content/uploads/2014/09/bryk-gomez_building-nics-education.pdf

Buck, G., Cook, K., Quigley, C., Eastwood, J., & Lucas, Y. (2009). Exploring how urban African-American girls position themselves in science learning: A sequential explanatory mixed-methods study. *Journal of Mixed Methods Research, 3*(4), 386–410.

Bush, S. B., Cox, R., & Cook, K. L. (2016). Building a prosthetic hand: Math matters. *Teaching Children Mathematics, 23*(2), 110–114.

Bush, S. B., & Cook, K. L. (2019). *Step into STEAM: Your standards-based action plan for deepening mathematics and science learning*. Thousand Oaks, CA: Corwin.

Bybee, R. (2010). Advancing STEM education: A 2020 vision. *Technology and EngineeringTeacher, 70*(1), 30–35.

Cook, K., Bush, S. B., & Cox, R. (2015). Engineering encounters: Creating a prosthetic hand. *Science and Children, 53*(4), 65–71.

Cook, K. L., Bush, S. B., & Cox, R. (2017). Engineering encounters: From STEM to STEAM. *Science and Children, 54*(6), 86–93.

Council of Chief State School Officers. (2010). *Common Core State Standards Initiative: Mathematics standards*. Washington, DC: National Governors Association and Author. Retrieved from http://www.corestandards.org/Math/

Cox, R., Hunter, K., Cook, K. L., & Bush, S. B. (in press). Teaching STEAM through a problem-based paleontology exploration. *Science and Children*.

Czerniak, C. M. (2007). Interdisciplinary science teaching. In S. K. Abell & N. G. Lederman (Eds.), *Handbook of research on science education* (pp. 537–560). New York: Routledge.

Delaney, M. (2014). *Schools shift from STEM to STEAM*. Edtech, 2 April, 1–4. [Online]. Available from: http://www.edtechmagazine.com/k12/article/2014/04/schools-shift-stem-steam.

Desimone, L. M. (2009). Improving impact studies of teachers' professional development: Toward better conceptualizations and measures. *Educational Researcher, 38*, 181–199.

Early Childhood STEM Working Group. (2017). *Early STEM matters: Providing high quality STEM experiences for all young learners*. A Policy Report. Retrieved from http://ecstem.uchicago.edu. UChicago STEM Education.

Greeno, J. G., Collins, A. M., & Resnick, L. B. (1996). Cognition and learning. In D. Berliner & R. Calfee (Eds.), *Handbook of educational psychology* (pp. 15–46). New York: Macmillan.

Herro, D., & Quigley, C. (2016). Exploring teachers' perceptions of STEAM teaching through professional development: Implications for teacher educators. *Professional Development in Education*, 1–23. https://doi.org/10.1080/19415257.2016.1205507

Hmelo-Silver, C. E. (2004). Problem-based learning: What and how do students learn? *Educational Psychology Review, 16*, 235–266.

Hom, E. J. (2014). *What is STEM education?* [Website]. http://www.livescience.com/43296-what-is-stem-education.html

Hunter, K., Cox, R., Bush, S. B., Cook, K. L., & Jamner, J. (2017). A paleontology investigation: "Unearthing" the mathematics. *Teaching Children Mathematics, 23*(7), 438–441.

Hurley, M. M. (2001). Reviewing integrated science and mathematics: The search for evidence and definitions from new perspectives. *Reviewing Integrated Science and Mathematics, 10*(5), 259–268.

Institute of Design at Stanford. (2016). Retrieved from http://dschool.stanford.edu/

Klein, J. T. (2014). Interdisciplinarity and transdisciplinarity: Keyword meanings for collaboration science and translational medicine. *Journal of Translational Medicine and Epidemiology, 2*(2), 1024.

Lave, J., & Wenger, E. (1991). *Situated learning: Legitimate peripheral participation.* Cambridge, UK: Cambridge University Press.

Lewis, S., Alacaci, C., O'Brien, G., & Jiang, Z. (2002). Preservice elementary teachers' use of mathematics in a project-based science approach. *School Science and Mathematics, 102*(4), 172–180. https://doi.org/10.1111/j.1949-8594.2002.tb18199.x

Loucks-Horsley, S., Stiles, E., Mundry, S., Love, N., & Hewson, P. (2010). *Designing professional development for teachers of science and mathematics* (3rd ed.). Thousand Oaks, CA: Corwin.

McAleer, S. D. (2008). Professional growth through mentoring: A study of experienced mathematics teachers participating in a content-based online mentoring and induction program. *Dissertation Abstracts International-A, 69*(08). (UMI No. 3319930).

NGSS Lead States. (2013). *Next generation science standards: For states, by states.* Washington, DC: National Academies Press. Retrieved from http://www.nap.edu/catalog/18290/next-generation-science-standards-for-states-by-states

Ostler, E. (2012). 21st century STEM education: A tactical model for long-range success. *International Journal of Applied, 2*(1), 28–33.

Pang, J., & Good, R. (2000). A review of the integration of science and mathematics: Implications for further research. *School Science and Mathematics, 100*(2), 73–82.

Park Rogers, M. A., & Abell, S. K. (2008). The design, enactment, and experience of inquiry-based instruction in undergraduate science education: A case study. *Science Education, 92*(4), 591–607.

Peppler, K. (2013). STEAM-powered computing education: Using e-textiles to integrate the arts and STEM. *Computer, 46*, 38–43. https://doi.org/10.1109/MC.2013.257

Putnam, R. T., & Borko, H. (2000). What do new views of knowledge and thinking have to say about research on teacher learning? *Educational Researcher, 29*, 4–15.

Read, T. (2013). *STEM can lead the way: Rethinking teacher preparation and policy.* California STEM Learning Network. Retrieved from http://face.ucr.edu/STEM%20Can%20Lead%20the%20Way%20Executive%20Summary.pdf

Savery, J. R. (2006). Overview of problem-based learning: Definitions and distinctions. *Interdisciplinary Journal of Problem-Based Learning, 1*(1), 9–20.

Smith, C. E., & Paré, J. N. (2016). Exploring Klein bottles through pottery: A STEAM investigation. *The Mathematics Teacher, 110*, 208–214.

Sztajn, P. (2011). Research commentary: Standards for reporting mathematics professional development in research studies. *Journal for Research in Mathematics Education, 42*, 220–236.

Timperley, H. S. (2011). *Realizing the power of professional learning.* Maidenhead, UK: Open University Press.

Watanabe, T., & Huntley, M. A. (1998). Connecting mathematics and science in undergraduate teacher education programs: Faculty voices from the Maryland collaborative for teacher preparation. *School Science and Mathematics, 98*(1), 19–25.

Wynn, T., & Harris, J. (2012). Toward a STEM + arts curriculum: Creating the teacher team. *Art Education, 65*(5), 42–47.

Chapter 3
Creating a STEAM Map: A Content Analysis of Visual Art Practices in STEAM Education

Christine Liao

STEAM Is Hot

STEAM is related to the STEM movement in education, itself a much-discussed topic in the pre-K-12, college, and community education contexts in the past few years. From policy makers to administrators, educators, and the media (e.g., Bertram, 2014; Jolly, 2014; Krigman, 2014; Pomeroy, 2012), the term STEM is used in educational contexts to refer to educational approaches and practices designed to encourage students to participate in STEM fields. Adding the arts to the STEM acronym, according to some advocators, emphasize the need for educating students to become innovators capable of competing in the global economy (Eger, 2013; Maeda, 2012). Professionals in the field of art education likewise have a stake in contributing to outcomes of educating creative and innovative students. It is, therefore, important that art educators understand what the term STEAM means in the context of their curricula.

The National Art Education Association (NAEA) sets out a position statement on STEAM education in April 2014, which defines the STEAM approach as "the infusion of art and design principles, concepts, and techniques into STEM instruction and learning" (National Art Education Association, 2014). This definition indicates that STEAM refers to the integration of any art and design learning into STEM; yet, instead of seeing STEAM as more than an instructional approach, this definition refers only to STEM instruction with/through art. The position statement also emphasizes collaboration among educators from different fields. In addition, it provided a resource detailing five other definitions of STEAM,[1] thereby indicating the organization's recognition of the term's complexity. The position statement is a

---

[1] See http://www.arteducators.org/research/STEAM_Definitions_Document.pdf

---

C. Liao (✉)
University of North Carolina Wilmington, Wilmington, NC, USA

© Springer Nature Switzerland AG 2019
M. S. Khine, S. Areepattamannil (eds.), *STEAM Education*,
https://doi.org/10.1007/978-3-030-04003-1_3

good start regarding helping art educators understand STEAM teaching, but other tools that can provide a range of perspectives on STEAM education are also needed.

My purpose for researching STEAM practices is to provide ways for art educators both to see their STEAM practices/curriculum in perspective and to think beyond the limitations of STEAM. I began my inquiry with these questions: What do current STEAM education practices look like? What are the goals of these practices? And, how do these practices relate to each other and the spectrum of art education and STEM Education fields?

I will start by discussing what STEAM education is and by reviewing a range of approaches to it to provide a foundation based on which educators can make their own inquiries into and create their own practices about STEAM. Based on my analysis of existing discussions focused on STEAM curricula and projects/lessons, I created the STEAM map presented in the current study to locate projects in relation to their emphasis on specific elements and goals. The STEAM map can be seen as visualized data of published current STEAM practices. Although the map does not include all STEAM curriculum practices and certainly does not extend to all the related possibilities in this area, it locates the preponderance of the practices in use and offers pathways for imagining new horizons. Overall, the map provides ways for educators to capitalize on STEAM practices, to reflect on the limitations of these, and to imagine and implement new approaches.

STEAM Curriculum Approaches

Despite the fact that STEM education has been around since the 1990s (Sanders, 2008), the usage of the term STEM is vague (Angier, 2010). A discussion of STEM in this regard is beyond the scope of this article; nonetheless, in education, STEM is considered as education in the field of science, technology, engineering, and math. However, approaches to teach students in learning these subjects and pursuing a career in these areas vary. Subsequently, adding the arts to STEM makes the usage of the term STEAM also vague. STEAM is generally understood to be a curriculum approach based on the STEM curriculum, i.e., science, technology, engineering, and math as the core subjects but with the arts (A) integrated either to the core subjects or to cultivate students' creativity and lead to innovation in STEM fields. The A includes helping students' learning of not only the visual arts but also the performing arts including music, drama/theater, and dance; and some researchers (e.g., Lewis, 2015) also advocate the inclusion of liberal arts subjects. In this article, I highlight only the integration of visual arts and design in order to focus on the implementation of STEAM in the field of art education. However, this does not mean that the other arts areas are excluded from my operational definition of the term.

Stroksdieck (2011) contends that there are two prevailing arguments in support of implementing STEAM education. According to one argument, art provides ways of seeing and knowing that differ from those captured by STEM and is, therefore, a valuable tool for apprehending STEM subjects. Stroksdieck provides data visualization as an example for this argument. Given that art can help learners to develop an understanding of scientific concepts (Dhanapal, Kanapathy, & Mastan, 2014), visualization is seen as a way to connect art and science such that it could constitute a STEAM practice in art education (Knochel, 2013). Knochel argues that art education can contribute to the understanding and discussion of how image manipulations of science data represent the "truth."

The other argument focuses on teaching art as a way to cultivate creative people (Stroksdieck, 2011). According to Trilling and Fadel (2009), even though STEM skills are in demand because they are considered necessary for careers in the twenty-first century, creativity will eventually be seen as even more important. Therefore, Trilling and Fadel proposed integrating the arts into STEM to create a STEAM focus as a central goal for twenty-first-century education. In accord with this argument, STEAM is seen as beneficial to students and to economic growth (Hanushek, Jamison, Jamison, & Woessmann, 2008). For example, John Maeda, a former president of the Rhode Island School of Design, is championing the notion of changing STEM to STEAM from exactly this viewpoint (Maeda, 2012, 2013; "STEM to STEAM," n.d.). Maeda views STEAM from the perspective of art and design education, arguing that an emphasis on STEM alone is not enough for the needs of twenty-first-century innovation. He advocates for art and science subjects to be taught together. For Maeda, the A in STEAM stands principally for visual art and design. In this view, the purpose of STEAM is to educate people to use the skills and creativity cultivated in art and design along with STEM knowledge in order to innovate and contribute to the progress of society. STEM is, therefore, of fundamental importance to social progress and economic growth, and art is the key to the innovation necessary for this kind of advancement ("STEM to STEAM" n.d.; White, 2010).

Design Education-Based STEAM Approach

Maeda's (2013) idea of emphasizing the development of design skills and creativity in an effort to create innovators is well received by art and design educators. Advocates for art and design education generally see STEAM as a way to highlight the importance of design education (Bequette & Bequette, 2012; Watson, 2015). Design education, argued Vande Zande (2010), would broaden the aims of art education and connect the economic goal that is not often present in the fine arts. Therefore, design can act as the bridge between STEM and the arts. This approach situates design education as a fundamental kind of STEAM education.

Design Thinking, typically understood as a creative problem-solving process, is often the method used to approach STEAM from the perspective of design education (Bequette & Bequette, 2012; Bush, Cox, & Cook, 2016; Gross & Gross, 2016). Researching what design thinking is and how it is taught, Donar (2011) found that "more integrated, cross–disciplinary, and holistic approach" (p. 98) is the trend of design thinking in education. Along with positioning design thinking as interdisciplinary (Cross, 2011), this inter/cross-disciplinary nature of design thinking process creates the connection between design and STEM. This perspective recognizes the complexity of the design thinking process and its potential to teach students to become critical thinkers equip with twenty-first-century skills needed across disciplines and careers (Watson, 2015).

Collaborative Approach to STEAM

For art and design educators, STEAM is not only about teaching design skills to advance STEM, but it should also be a collaborative effort between educators from various disciplines (Watson, 2016). Art and design educators, therefore, should communicate with others in STEM fields in order to determine how art can fit into the STEM/STEAM curriculum (Bequette & Bequette, 2012), and art teachers and STEM teachers should learn from each other (Wynn & Harris, 2012). This view constitutes the interdisciplinary and collaborative approach to STEAM. The concept is extended to taking the communication between art and STEM educators into a transdisciplinary space (Guyotte, Sochacka, Costantino, Walther, & Kellam, 2014). It is important to note that transdisciplinary is different from interdisciplinary. Nicolescu (1997) defines the goal of transdisciplinary (research) as "the understanding of the present world, which cannot be accomplished in the framework of disciplinary research" (para. 8). In other words, it is a space beyond (multi)disciplines. Guyotte et al. (2014) argue that a STEAM curriculum in the context of a transdisciplinary space can become a form of social practice. They also propose that working in a transdisciplinary STEAM space can help students develop creativity collaboratively (Guyotte, Sochacka, Costantino, Kellam, & Walther, 2015). At the heart of this position on STEAM is the practice of collaboration.

Collaboration in STEAM usually means that teachers/students from different disciplines work together. However, individual disciplines taught as discrete areas of focus still constitute the foundation of this approach. In other words, collaboration is based on the idea of benefiting from each other's differences. On a similar path, Yakman (2008, 2012) proposes the term STEAM written as $ST\sum@M^2$ from an integrated curriculum perspective. In her account, an integrated curriculum can break the discipline-based limitations of most established educational approaches. $ST\sum@M$ constitutes a curriculum framework that Yakman created to structure traditional academic subjects to be taught in an integrated curriculum (2008). She defines STEAM as "Science and Technology, interpreted through Engineering and the Arts, all based in a language of Mathematics" (2008, p. 21). In this view, the A in STEAM stands for all the arts as well as for the humanities and social sciences. This perspective focuses on a holistic approach to the curriculum (Yakman & Lee,

---

[2] The symbol $\sum$ (sigma), used in math to refer to the sum of all values, is used here to emphasize the STEM connection.

2012), such that Yakman advocates for the importance of all the subjects and, therefore, that students should gain a background in all of them. However, in this framework—referred to as the STEAM Pyramid[3]—the subjects are not of equal importance: The arts, humanities, and social sciences comprise only one-fifth of an integrated curriculum that remains focused on STEM subjects. Yakman advocates integrating different subjects through a common theme or project, and each individual subjects in this STEAM curriculum can be taught by different teachers to reduce the burden of teachers (Joe, 2017). In addition to that this approach still advocates a divide of individual subjects, this approach has evolved into a curriculum system that is taught to teachers through their specific professional development partnership and membership. Thus, the practice of this idea is not transparent to people who do not have access to their professional development and curriculum. It is unclear that how this idea is actually practiced in the day-to-day school curriculum.

Approaching STEAM Through Arts Integration

Perhaps, the practical approach to STEAM education is through arts integration. Advocates of arts integration are increasingly using the term STEAM as a vehicle for disseminating practices associated with arts integration. Unlike Yakman's integrated STEAM curriculum framework, arts integration is often discussed at the level of instructional approach and lessons, although its larger goal is also "integration," which can be implemented in a variety of ways. As such, much as there are many definitions of arts integration, a general understanding is that the term refers to the teaching of other subjects through the arts (Goldberg, 2011). Different terms such as arts infusion (McDonald, 2010) or arts-based teaching and learning (Marshall, 2014) are used to refer to arts integration. The Kennedy Center's definition of arts integration, as set out by Silverstein and Layne, focuses on the potential of using the arts to create the understanding of both the art subject and the subject(s) with which it is connected in a given curriculum:

> Arts integration is an approach to teaching in which students construct and demonstrate understanding through an arts form. Students engage in a creative process which connects an art form and another subject area and meets evolving objectives in both. (Silverstein & Layne, 2010)

Marshall (2014) discusses that Silverstein and Layne's definition of arts integration is multi-model arts-based learning, and this definition has gone beyond many other views that the arts production is only a strategy for teaching and learning other subjects. To further conceptualize arts integration, Marshall argues that arts integration is a transdisciplinary field and its pedagogy goes beyond disciplinary boundaries.

---

[3] The STEAM Pyramid can be found here: https://steamedu.com/pyramidhistory/

Viewed through the lens of arts integration, STEAM can be seen as a teaching approach focused on the benefits of integrated learning. However, no matter based on which perspective or definition of arts integration, embedding the A in STEM makes obvious that the purpose of STEAM is to integrate the arts into the teaching of STEM subjects and/or to teach STEM subjects through the arts. Even, as described in the Kennedy Center's definition, art integration should focus on fulfilling objectives for both the arts and another subject(s), emphasizing STEM subjects often results in the arts serve as knowledge delivering tool. In practice, the arts are often diluted in arts integration practices in the general classroom such that the focus is only the "main" subject(s) (LaJevic, 2013). Further, in Riley's (2013) view, "too often, the arts are used as enhancement in the lesson (think 'shadow boxes') rather than as a true means of connecting and communicating understanding" (para. 8). The mistreatment of the arts in art integration practice is also brought into the STEAM curriculum. Art is often reduced to only the aesthetics of a project. For example, in a fourth-grade STEAM project teaching students to create a prosthetic, the art learning in this project is described as "aesthetics of designing a prosthetic" (Bush et al., 2016, p. 111).

Although advocates argue that significant benefits can accrue from integrating the arts into STEM learning, as described by Catchen and DeCristofano (2015), many in STEM fields think of the arts as lacking rigor. Such a view of the arts on the part of STEM practitioners may have resulted in an apparent reluctance to use the term STEAM and a tendency to criticize this concept for drawing attention away from STEM (Dunning, 2013; May, 2015). Even though art educators know that students develop critical thinking and creative problem-solving skills through the arts (e.g., Eisner, 2002), misconceptions of the arts hinder efforts to bridge the gaps between those in STEM and those in non-STEM fields.

The NAEA's position statement falls within the view of STEAM associated with arts integration practices. However, educators should be cautious in regard to discussing STEAM solely through the lens of arts integration because the arts could come to be seen not as a subject area in its own right but merely as a tool for learning STEM. Even though advocates in this camp emphasize the importance of the arts, they are clearly a vehicle for teaching STEM subjects—not the other way around. Equating arts integration with STEAM or using the term STEAM to attract attention to arts integration could also result in a loss of the idea that all subjects should be treated as equal within the purview of arts integration.

Approaching STEAM Through Project-Based Learning and the Maker Movement

Whether STEAM is approached with an emphasis on collaboration or on integration, project-based learning (PBL), which is rooted in John Dewey's philosophy of learning by doing (Boss, 2011), is one of the most popular STEM/STEAM-teaching

3 Creating a STEAM Map: A Content Analysis of Visual Art Practices in STEAM... 43

approaches (Markham, 2012; Miller, 2014). According to Thomas (2000), project-based learning "involve[s] students in design, problem-solving, decision making, or investigative activities; give[s] students the opportunity to work relatively autonomously over extended periods of time; and culminate[s] in realistic products or presentations" (p. 1). PBL is also advocated by Yakman (2008) and by Bequette and Bequette (2012). Through processes associated with PBL, students can engage in learning and using STEAM knowledge and skills, and there is clear potential for art and STEM to be treated as equally important to the success of any given project.

In emphasizing learning through "doing/making" projects, however, many STEM/STEAM curricula confuse PBL with project-oriented learning (Larmer & Mergendoller, 2010; Robin, 2011). Project-based learning focuses on the problem-solving process, but project-oriented learning focuses on the finished product, which is often a craft item or school-style art (Gude, 2013). Although proponents for the arts emphasize "making" as an essential art skill, often very little, if any, art knowledge is taught in the process of creating these "projects." In an approach of this nature, hands-on making is the only way in which art is taught in STEAM. Consequently, art is not considered beyond its function as a tool/medium for STEM, or worse, its function is entirely decorative.

In another trend related to the rise of STEM and STEAM education, the maker movement also considers hands-on and making-centered approach (Dougherty, 2012), especially with digital media, electronic technology, and robotic technology. This movement has gained popularity among the general public and also made an impact on education (Halverson & Sheridan, 2014). In fact, some educators see the maker movement as an opportunity for STEM/STEAM education (Bevan, Gutwill, Petrich, & Wilkinson, 2015; Gerstein, 2013; Peppler & Bender, 2013). Interest in connecting the maker movement with STEAM education is often associated with the project-orientated and hands-on learning approach described above, but not necessary with PBL. As art is generally seen as a hands-on subject, art educators are also embracing the maker movement (Ciampaglia & Richardson, 2014). The critical pedagogy approach practiced by Ciampaglia (2014) is an important direction to consider STEAM education through maker movement.

However, this movement is often criticized for its white male-dominated culture (Grenzfurthner & Schneider, n.d.). Further, although art skills and creativity are valued, most of the creative "projects" are function-driven and, again, geared toward socially constructed male interests, such as electronics or robots (Buechley, 2013). Conceptual expression and cultural connections are rarely seen in this kind of projects. It is, therefore, important to identify what art can bring to the maker movement and to distinguish maker projects, such as assembling a robot, from more creative uses of such technology to express ideas. Some art educators' advocates for critical making (Patton & Knochel, 2017) aim at bringing maker movement to the next level. Similarly, Ciampaglia and Richardson (2014) emphasized using critical pedagogy to teach digital making. These examples are ways to consider a STEAM education beyond only creating high-tech craft.

Summary: Many Shades of STEAM

These different viewpoints on what constitutes STEAM education do not have clear boundaries. There is no single definition of STEAM or by extension of what constitutes STEAM practices. Instead, there is a broad spectrum of practices in the field of art education and arts integration all claiming the term STEAM.

Some practices referred to as STEAM focus exclusively on only one STEM subject in concert with one or more arts subjects. These can be seen as one piece of a STEAM curriculum, for example, creating digital media as a way of integrating technology literacy and art (Blair, 2015). As technology is one of the focal STEM subjects, a curriculum integrating digital technology/media creation would constitute a de facto STEM curriculum. The matter of whether any given curriculum of this kind is a STEAM curriculum depends on how art making is defined.

Other examples of STEAM education range from art projects with science as the central topic (e.g., Gibbons, 2015; Hare & Feierabend, 2015) to art projects with the focus on design (e.g., Giordano, 2015) and from lessons integrating art and math (e.g., Ward & Albritton, 2015) to lessons in which collaborative projects cross-disciplinary boundaries (e.g., Barnett & Smith, 2013) and to a focus on artists' creative explorations of STEM topics (e.g., Joksimovic-Ginn, 2014). Each of these practices has its own role and function at different levels of education.

In addition, artists' STEAM practices create another territory of STEAM practice for art educators to explore. Examples of STEAM practices in art can be seen in "Steam," a 2014 exhibition curated by Patricia Miranda at the ArtsWestchester's Arts Exchange. The exhibition included pieces showing the involvement of artists in scientific disciplines (Hodara, 2014). For example, artist Carl Van Brunt uses a fractal generator to create images of "mathematical nature." Another artist, William Meyer, created a transparent backpack with a "complex system of earth microorganisms, chemistry and botany" (Hodara, 2014, para. 7). In fact, artists have long worked with STEM knowledge, concepts, ideas, and technologies. Leonardo da Vinci would be a famous example. How artists' practices can be translated into art teaching and STEAM education is an important question for art educators.

Whatever the viewpoint and whatever the practices associated with STEAM, the question arises as to its value for contemporary society. Although STEM knowledge and skills are essential for success in the twenty-first century and the combination of art and STEM holds promise for improving the quality of life, STEM and STEAM are not the panacea. As some critics point out, a greater focus on STEM subjects means that there is less time for other areas, such as the humanities (Ossola, 2014; Zakaria, 2015). STEAM, however, can fill in some of the missing pieces for STEM education. But more importantly, educators need to see the bigger picture in regard to STEAM education practices to address the issues. Therefore, in order to see what current STEAM education practices look like, I created a snapshot of current practices and presented it as the STEAM map in the next section.

Creating a STEAM Map

Method

The focus of this study is to understand what the STEAM curriculum in art education and the integration of visual arts with STEM look like in practice. The result of the visualized data is presented as a STEAM map showing the distributions of different emphasizes of STEAM curricula. I conducted content analysis using magazine articles, lessons published online, and journal articles published from 2012 to 2016 in the field of art education and STEAM (Table 3.1) aims for practitioners to see the trends and approaches to STEAM curriculum. The data collection of this study focused solely on visual arts-related STEAM curricula, projects, lessons, and artwork. These published projects serve as examples for those interested in STEAM education.

Content analysis is traditionally used as a quantitative research method, but with the development of qualitative content analysis, it has become a method to bridge quantitative and qualitative research approaches (Duriau, Reger, & Pfarrer, 2007; Hsieh & Shannon, 2005; Mayring, 2000). In this research, my approach to content analysis is mixed method. I use qualitative content analysis, "a research method for the subjective interpretation of the content of text data through the systematic

Table 3.1 Sources of Data

Publication/online community	Description	Number of Projects Selected (2012–2016)
School Arts	This is a practitioner magazine in art education	12
Arts and Activities	This is a practitioner magazine in art education	1
Art Education	This is a practitioner journal in art education	8
EducationCloset	Although EducationCloset (http://educationcloset.com/steam/lessons/) is not a publication, it is a popular arts integration and STEAM online community that includes lesson plans	6
STEAMed Magazine	This is a practitioner magazine focused specifically on STEAM	13[a]
The STEAM Journal	This is a STEAM-focused journal that has multiple sections including Articles, Artwork, Field Notes, and Reflections	15
		Total: 55

[a]Due to the access issue, only the articles published in 2015 in the *STEAMed Magazine* are collected. The magazine's inauguration issue is January 2015

classification process of coding and identifying themes or patterns" (Hsieh & Shannon, 2005, p. 1278), and specifically using the inductive category development method (Mayring, 2000) to create category and analysis the text. The texts are interpreted based on the different goals and focus of various STEAM education approaches discussed earlier. I also borrow from the counting method of quantitative content analysis to present the trend of STEAM and create the STEAM map.

I only collected data on projects that use either the term STEAM or STEM in conjunction with art. The projects included range from early childhood to college to community contexts. In addition, even though some of the projects focused on only one STEM subject, they were included providing either the term STEAM or STEM and art was used. Some of the articles selected did not specifically focus on discussing a STEAM project but were selected if one or more STEAM practices were briefly mentioned as an illustration of the author's idea of STEAM curriculum. On the other hand, some articles, even though focusing on discussing STEAM, were not selected, if no specific STEAM art-making curriculum project was included. In regard to the data collected, discussions of school-wide STEAM curricula that did not mention a specific project were not included. All the STEAM projects from which data were collected met the following criteria:

1. Included the visual arts
2. Used the term STEAM or STEM and art or were published in a STEAM-focused publication
3. Involved creative production (creating an artwork or a STEM application), including through participatory art

Although the current study does not present a comprehensive survey, the publications and the website included are representative places in which STEAM curricula are published and can be easily accessed by practitioners, and these are places advocate for STEAM practices in art education and other disciplinary areas.

Fifty-five STEAM projects involving the visual arts are included in my analysis. These projects are analyzed using categories created through qualitative content analysis inductive development method. This method allows me to create categories through a series of steps starting from determining the categories and adjusting the categories base on the materials read. I started with listing possible categories based on the main approaches to STEAM practice in the projects. Several categories were created in order to determine the different objectives/approaches distinguishing the various projects. After setting up the initial categories, the data texts are carefully read and interpreted, and some categories are divided into smaller categories or combined into one category. After reading most of the materials, the categories are adjusted again. The final analysis is based on the categories that have been revised a few times (Mayring, 2000). Some of the projects selected utilize multiple approaches. I categorize these projects only based on the main approach and goal.

After creating the categories (Table 3.2), the data are further analyzed based on the following dimensions: art, STEM, creative expression/application, and knowledge/skills. The dimensions are created based on the traditional disciplinary concept separating art and STEM and the learning objectives from revised Bloom's

3 Creating a STEAM Map: A Content Analysis of Visual Art Practices in STEAM... 47

Table 3.2 Categories

A	Creating art informed by STEM or art (concept) based on STEM knowledge
B	Using STEM knowledge to create art (STEM knowledge is a tool to achieve art-making goals)
B-1	Creating art while also acquiring STEM knowledge driving by the art/medium and process
B-2	Creating art with STEM knowledge to educate/advocate STEM-related or other issues
C	Creating art and representing researched/learned STEM knowledge in creative ways
C-2	Learning design process (Design Thinking) as a way to learn STEAM skill/knowledge
E	Using art and design skills and STEM knowledge to create STEM applications (projects)
F	Creating real-world application-oriented projects while learning both art and STEM content
G	Creating art to learn, understand, and represent/demonstrate STEM content
G-1	Using art techniques and exploring art medium to learn STEM skills/knowledge/content
G-2	Using art to illustrate STEM content/data visualization
H	Learning a skill/concept that is shared knowledge/skill between art and STEM
I	Engaging in hands-on activities to create a STEM/STEAM project (in makerspace setting)
Y	Collaborating with people in STEM to create art to highlight/solve STEM issues
Z-1	Collaborating with (teaching) artists to create art that integrates with STEM content/knowledge or collaborating in an interdisciplinary way to create art while learning both STEM and art
Z-2	Exploring a topic/theme/concept through interdisciplinary learning or collaboration
Z-3	Solving a problem through transdisciplinary collaboration

taxonomy (Anderson & Krathwohl, 2001). The art or STEM dimensions are based on the project's focal discipline. The creative expression/application and knowledge/skill learning dimensions are based on the main learning objective of the project. Although most of the projects have goals relating to both art and STEM, many give considerably more weight to one or the other of these. The categories are located on the map based on the four dimensions outlined.

Drawing the Map

In order to visualize the distribution of the STEAM practices, I created a map using a coordinate system that can represent the data in two-dimensional space. Using the quantitative method of counting, the STEAM map shows the number of STEAM projects found in the collected data and the location of the categories. The location of each category was determined by its relativeness to the different dimensions. The center area is where different dimensions meet. It indicates the integration and balance of different dimensions. Therefore, the center is circled and identified as an inter/transdisciplinary collaboration area. In order to verify that the locations of the categories are correctly represented on the map, the categories were grouped based

Table 3.3 STEAM Approaches and Categories

STEAM approaches	Categories
Shared knowledge	H
Design-based STEAM education	C-2
Collaboration, inter/transdisciplinary	Y, Z-1, Z-2, Z-3
Arts integration approach	B-1, C, G, G-1
Project-based learning	B-2, E, F
Maker movement-based approach	I
Artist and art education focused	A, B

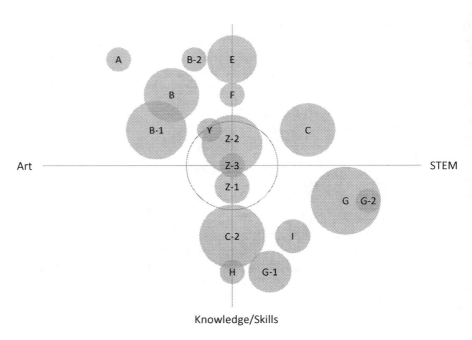

Fig. 3.1 STEAM Map

on the approaches discussed earlier (Table 3.3), which include focuses on shared knowledge, design education or design thinking approach, arts integration approach, collaboration and inter/transdisciplinary approach, project-based learning, maker movement-based approach, and artists and art-making focused approach. The groups allowed me to see that most categories within the same STEAM approach are located close to each other, with the exception of arts integration approach. This is due to the complex nature of this approach. The visualized data was created using the Excel program. The number of projects in each category is indicated by a circle of a corresponding size: The more projects in a category, the bigger the circle used to represent that category. In Table 3.2, a complete list of the categories is presented. These are the keys for reading the STEAM map (Fig. 3.1).

Stories the Map Tells

The map is by no means comprehensive and is limited in some ways. However, it does provide a way to visualize the different approaches and objectives between current STEAM education practices. As I collected only data about visual arts, this study does not represent other arts areas that are included in STEAM. The projects collected ranged widely from single STEAM lessons/projects to collaborations across disciplines and beyond schools to community education.

The inter- or transdisciplinary collaborative approaches are located in the middle of the map (represented by Y, Z-1, Z-2, and Z-3) because these projects are not situated specifically in the context of either the visual arts or of a STEM subject and are not focused on merely learning a basic knowledge/skill or creative conceptual expression. That is, inter- or transdisciplinary collaborative approaches usually bring both learning/applying knowledge and creative expression together to achieve a goal, such as solving a problem. For example, the *EcoScience + Art* initiative led by Changwoo Ahn (2015) is a collaborative STEAM initiative that is continuing to evolve. One project in this initiative, *The Rain Project*, was designed to "promote participation and collaboration in the context of ecological literacy and campus sustainability" (Ahn, 2015, p. 4). Students from various disciplines worked together to build a floating wetland over the period of a year in order to address stormwater issues and improve the quality of the water at their campus. This project is an inter-disciplinary collaboration that led to real-world problem solving.

A general trend captured by the STEAM map is represented by the categories with the larger circles (G, B-1, C-2). These locations are away from the center of the map, indicating that many of the practices in these contexts are still situated in specific disciplines. Further, more than half of the projects in these categories focus more on learning the basic art or STEM knowledge/skills than on creative applications.

The map shows that two kinds of STEAM practices currently dominate the discourse in the field of art education: practices that proceed from the idea of using STEM knowledge to create art represented by the B, B-1, and B-2 categories, and practices that proceed from the idea of creating art to learn, understand, represent, or demonstrate STEM content, represented by the G category (the largest circle on the map). The G and B-1 categories belong to arts integration approach, which uses art creation as a way to learn STEM content. The difference between these two categories is that projects in G category use art making as a way to understand STEM content or demonstrate learning. The goal is STEM content learning, and art is the vehicle to achieve the goal. While the projects in B-1 category focus on art making but use the opportunity to also learn about STEM content related to the art medium or making process. The goal of these projects is art learning oriented. The G category is strongly corresponding with the goal of arts integration set out by Kennedy Center, which is to understand and demonstrate learning of other subjects through art making. The larger number of projects in this category indicated that in the field of art education, arts integration is a popular approach to teaching STEAM. The C category is similar to G category, but the difference is that C category use art to represent learning in STEM area, but the art making and the STEM learning are likely separated. While the STEM learning in G category is through the art making, so the learning is integrated. In addition, the C-2 category is another popular STEAM practice. It represents the design education-based approach to STEAM education. The projects in this category teach students to use design thinking process to design their project and learn about art and STEM together.

Many projects are situated in the categories located along the X axis, which is in the middle of art and STEM. This means that these projects view art and STEM equally and combine both areas to reach the maximum benefit. It is interesting to note that there is no category locates at the lower left section, which indicates learning basic art skills. This might be due to the limitation of selected data. In addition, the center area with categories Z-1, Z-2, and Z-3 is showing strong numbers of advocates and projects. This is also a good indication that the collaborative approach is practical, and many educators see the value of this approach.

Beyond the Map

The map shows only the approaches taken in the projects collected. Therefore, there are unmapped territories. The STEAM curriculum projects collected in this study are the ones that have been published and publicly available. Thus, some popular curriculum practices, such as those developed in Yakman's STEAM Education professional development program are not included. In addition, the map does not show school-wide STEAM curriculum reform. It would be a worthwhile endeavor to consider how specific STEAM-focused schools approach STEAM curricula in the future. Further, it is important to note that there are some STEAM practices, although relatively few, in the STEM field that were not included in this study because I intended to focus on how the art fields (art education and arts integration) practices STEAM education. Despite its limitations, however, the map provides a useful tool based on which art educators can locate their STEAM projects in the context of current STEAM education practices. Art educators can look at the dimensions and locations of each category and decide how to adjust their STEAM curriculum/project to considering expanding the curriculum/project into more dimensions.

I would like to propose more dimensions for a future map as well. As the STEAM map shows, most of the projects center on either learning art or STEM knowledge/skills or creating an art/functional application project. A very small number of

3 Creating a STEAM Map: A Content Analysis of Visual Art Practices in STEAM…

projects focus on social practices, justice, or environmental awareness (e.g., Ahn, 2015; Fontes, 2015; Guyotte et al., 2014). Aligned with Patton and Knoche'sl (2017) view of the importance of criticality of digital making in STEAM, in my view, a STEAM education that begins from and engages with a conceptual exploration of our society pertaining to issues such as social and economic justice, environmental sustainability, or human rights has the potential to be truly transformative. When there is sufficient STEAM curriculum to warrant including dimensions for these areas of concern, it will mean that steps have been taken to enrich the STEAM curriculum and prepare young children and adults to better understand and play a positive role in the society they live in as a whole.

Future Directions of STEAM Education

The idea that STEAM education has the ability to transform students' learning of STEM subjects and can provide them with important skills needed for the future success has driven the discourse on STEAM overall. Art education is in a unique position in regard to the STEAM movement because of the shared knowledge between art and STEM subjects in many respects, such as patterns in both art and math. However, it is also necessary to attend to the subject areas and discourses that the STEAM movement neglects. As there are many perspectives and approaches on STEAM, some worried that the importance of liberal arts education might lose in the current. Therefore some commentators are advocating for including the liberal arts under A in STEAM (e.g., Gogus, 2015; Lewis, 2015). Further, no matter STEM or STEAM movement, the perception that the STEM fields are driving social progress and that such progress is good for society should be carefully examined. STEAM education should also include education focused on our relationship with and the ethics of new technology and scientific advances. For example, although it is at the center of efforts to merge science and art, BioArt has long been considered controversial in terms of its ethics (Munster, 2005; Zylinska, 2009). From a diversity and gender equality perspective, we want to encourage girls to participate in STEM fields, but at the same time, the working conditions in these fields for women need to be improved (Fouad & Singh, 2011; Steele, 2013). In addition, some argue that including the arts in STEM will attract girls to STEM fields (Gustlin, 2014; Santana, 2015). However, doing so might bolster stereotypical ideas about girls' preferences. The maker movement, which is another strong force pushing the STEAM movement, has been criticized for its white male-centered culture (Chachra, 2015). These are all issues that need to be addressed in the context of STEAM.

In this context, the question is this: How can art teachers teach STEAM? The answer: Art making, especially involving conceptual exploration, can function well as the central approach. Art educators who advocate for STEAM consider art making to be an important component of the STEAM curriculum (Bequette & Bequette, 2012; Wynn & Harris, 2012). Although working in a transdisciplinary space is the ideal, it requires teamwork and extended planning. In reality, this would be a big step forward for many educators. Working toward the goal of establishing a trans-disciplinary space, individual teachers can also start with small strategies and class projects in order to start building bridges between art and STEM.

The STEAM map should serve as a way to locate the goals and approaches of current STEAM practices and as a basis for art educators and other stakeholders to envision advancing into other areas. Although the study is limited, it shows that educators are exploring what constitutes as STEAM curriculum in the past few years through different approaches. In the flow of discussions on this subject, STEAM functions as a political term by bringing attention to the arts. I encourage more art educators to provide their perspectives in an effort to shape the directions that STEAM takes.

References

Ahn, C. (2015). EcoScience + Art initiative: Designing a new paradigm for college education, scholarship, and service. *STEAM Journal, 2*(1), Article 11. https://doi.org/10.5642/steam.20150201.11.

Anderson, L. W., & Krathwohl, D. R. (Eds.). (2001). *A taxonomy for learning, teaching, and assessing: A revision of Bloom's taxonomy of educational objectives*. New York: Longman.

Angier, N. (2010, October 5). STEM education has little to do with flowers. *New York Times*. p. D2.

Barnett, H., & Smith, J. R. A. (2013). Broad vision: The art & science of looking. *The STEAM Journal, 1*(1), Article 21. https://doi.org/10.5642/steam.201301.21.

Bequette, J. W., & Bequette, M. B. (2012). A place for art and design education in the STEM conversation. *Art Education, 65*(2), 40–47.

Bertram, V. (2014). STEM or STEAM? We're missing the point. Retrieved from http://www.huffingtonpost.com/vince-bertram/stem-of-steam-were-missin_b_5031895.html

Bevan, B., Gutwill, J. P., Petrich, M., & Wilkinson, K. (2015). Learning through STEM-rich tinkering: Findings from a jointly negotiated research project taken up in practice. *Science Education, 99*(1), 98–120.

Blair, J. M. (2015). BioShocked: environmental game design. *School Arts, 114*(8), 34–35.

Boss, S. (2011, September 20). Project-based learning: A short history. Retrieved from http://www.edutopia.org/project-based-learning-history

Buechley, L. (2013, October). *Thinking about making*. Presented at the FabLearn Conference, Palo Alto, CA. Retrieved from http://edstream.stanford.edu/Video/Play/883b61dd951d4d3f90abee c65eead2911d

Bush, S. B., Cox, R., & Cook, K. L. (2016). A critical focus on the M in STEAM. *Teaching Children Mathematics, 23*(2), 110–114. https://doi.org/10.5951/teacchilmath.23.2.0110

Catchen, R. D., & DeCristofano, C. (2015). What's wrong with interpretive dance? Embracing the promise of integrating the arts into STEM learning. *The STEAM Journal, 2*(1), Article 9. https://doi.org/10.5642/steam.20150201.9.

Chachra, D. (2015, January 23). Why I am not a maker. Retrieved from http://www.theatlantic. com/technology/archive/2015/01/why-i-am-not-a-maker/384767/

Ciampaglia, S. (2014). Critical pedagogy 2.0: Researching the visual culture of marketing with teenage coresearchers. *Studies in Art Education, 51*(6), 359–369.

Ciampaglia, S., & Richardson, K. (2014). *The Plug-in studio: Art education for the maker age*. Presented at the FabLearn 2014, Stanford, CA: Stanford University. Retrieved from http://fablearn.stanford.edu/2014/wp-content/uploads/fl2014_submission_6.pdf

Cross, N. (2011). *Design thinking: Understanding how designers think and work*. New York: Berg.

3 Creating a STEAM Map: A Content Analysis of Visual Art Practices in STEAM...

Dhanapal, S., Kanapathy, R., & Mastan, J. (2014). A study to understand the role of visual arts in the teaching and learning of science. *Asia-Pacific Forum on Science Learning and Teaching, 12*(2), 12.

Donar, A. (2011). Thinking design and pedagogy: An examination of five Canadian post-secondary courses in design thinking. *Canadian Review of Art Education, 38*, 84–102.

Dougherty, D. (2012). The maker movement. *Innovations, 7*(3), 11–14.

Dunning, B. (2013). *Can we be clear on something? It's STEM, not STEAM.* Retrieved from http://www.skepticblog.org/2013/03/14/stem-not-steam/

Duriau, V. J., Reger, R. K., & Pfarrer, M. D. (2007). A content analysis of the content analysis literature in organization studies: Research themes, data sources, and methodological refinements. *Organizational Research Methods, 10*(1), 5–34. https://doi.org/10.1177/1094428106289252

Eger, J. (2013). STEAM...now! *STEAM Journal, 1*(1), Article 8. https://doi.org/10.5642/steam.201301.08.

Eisner, E. W. (2002). *The arts and the creation of mind.* New Haven, CT: Yale University Press.

Fontes, K. (2015, January). Art is awesome! Creating using LEGO animation. *STEAMed,* 22–23.

Fouad, N. A., & Singh, R. (2011). *STEMMING the tide: Why women leave engineering.* Milwaukee, WI: University of Wisconsin-Milwaukee. Retrieved from http://energy.gov/sites/prod/files/NSF_Stemming%20the%20Tide%20Why%20Women%20Leave%20Engineering.pdf

Gerstein, J. (2013, July 23). *STEAM and maker education: Inclusive, engaging, self-differentiating.* Retrieved from https://usergeneratededucation.wordpress.com/2013/07/23/steam-and-maker-education-inclusive-engaging-self-differentiating/

Gibbons, E. (2015). Expressive organic forms. *School Arts, 115*(1), 29–31.

Giordano, K. (2015). The physics of sitting. *School Arts, 115*(1), 20–21.

Gogus, A. (2015). Reconceptualizing liberal education in the 21st century: The role of emerging technologies and STEAM fields in liberal education. In X. Ge, D. Ifenthaler, & J. M. Spector (Eds.), *Emerging technologies for STEAM education: Full STEAM ahead* (pp. 277–292). New York: Springer.

Goldberg, M. (2011). *Arts integration: Teaching subject matter through the arts in multicultural settings* (4th ed.). Boston: Pearson.

Grenzfurthner, J., & Schneider, F. A. (n.d.). Hacking the spaces. Retrieved from http://www.monochrom.at/hacking-the-spaces/

Gross, K., & Gross, S. (2016). Transformation: Constructivism, design thinking, and elementary STEAM. *Art Education, 69*(6), 36–43.

Gude, O. (2013). New school art styles: The project of art education. *Art Education, 66*(1), 6–15.

Gustlin, D. (2014). *Why add art to STEAM?* Retrieved from http://educationcloset.com/steam/why-add-art-to-steam/

Guyotte, K. W., Sochacka, N. W., Costantino, T. E., Kellam, N. N., & Walther, J. (2015). Collaborative creativity in STEAM: Narratives of art education students' experiences in transdisciplinary spaces. *International Journal of Education and the Arts, 16*(15), 1–38.

Guyotte, K. W., Sochacka, N. W., Costantino, T. E., Walther, J., & Kellam, N. N. (2014). STEAM as social practice: Cultivating creativity in transdisciplinary spaces. *Art Education, 67*(6), 12–19.

Halverson, E. R., & Sheridan, K. M. (2014). The maker movement in education. *Harvard Educational Review, 84*(4), 495–504.

Hanushek, E. A., Jamison, D. T., Jamison, E. A., & Woessmann, L. (2008). Education and economic growth: It's not just going to school but learning that matters. *Education Next, 8*(2), 62–70.

Hare, T., & Feierabend, J. (2015). Wondrous watercolor cell illustrations. *School Arts, 115*(1), 36–37.

Hodara, S. (2014, July 25). Putting the A in STEAM. *New York Times.* Retrieved from http://www.nytimes.com/2014/07/27/nyregion/putting-the-a-in-steam.html

Hsieh, H.-F., & Shannon, S. E. (2005). Three approaches to qualitative content analysis. *Qualitative Health Research, 15*(9), 1277–1288. https://doi.org/10.1177/1049732305276687

Joe, F. (2017). STEAM teaching and learning. *China Science & Technology Education, 253,* 6–7.

Joksimovic-Ginn, B. (2014). Can art stress? *The STEAM Journal, 1*(2), Article 16. https://doi.org/10.5642/steam.20140102.16.

Jolly, A. (2014, November 18). STEM vs. STEAM: Do the arts belong? *Education Week: Teacher.* Retrieved from http://www.edweek.org/tm/articles/2014/11/18/ctq-jolly-stem-vs-steam.html

Knochel, A. (2013). Histochemical seeing: Scientific visualization and art education. *Studies in Art Education, 54*(2), 187–190.

Krigman, E. (2014, February 13). *Gaining STEAM: Teaching science through art.* Retrieved from http://www.usnews.com/news/stem-solutions/articles/2014/02/13/gaining-steam-teaching-science-though-art

LaJevic, L. (2013). Arts integration: What is really happening in the elementary classroom? *Journal for Learning Through the Arts, 9*(1), 1–28.

Larmer, J., & Mergendoller, J. R. (2010). *The main course not dessert.* Novato, CA: Buck Institute for Education. Retrieved from http://bie.org/object/document/main_course_not_dessert

Lewis, A. L. (2015). Putting the "H" in STEAM: Paradigms for modern liberal arts education. In X. Ge, D. Ifenthaler, & J. M. Spector (Eds.), *Emerging technologies for STEAM education: Full STEAM ahead* (pp. 259–276). New York: Springer.

Maeda, J. (2012, October 2). STEM to STEAM: Art in K-12 is key to building a strong economy. Retrieved January 17, 2015, from http://www.edutopia.org/blog/stem-to-steam-strengthens-economy-john-maeda

Maeda, J. (2013). STEM + Art = STEAM. *The STEAM Journal, 1*(1), Article 34. https://doi.org/10.5642/steam.201301.34.

Markham, T. (2012). *STEM, STEAM, and PBL.* Retrieved from http://edge.ascd.org/blogpost/stem-steam-and-pbl

Marshall, J. (2014). Transdisciplinarity and art integration: Toward a new understanding of art-based learning across the curriculum. *Studies in Art Education, 55*(2), 104–127.

May, G. S. (2015). *STEM, not STEAM.* Retrieved from https://www.insidehighered.com/views/2015/03/30/essay-criticizes-idea-adding-arts-push-stem-education

Mayring, P. (2000). Qualitative content analysis. *Forum: Qualitative Social Research, 1*(2). Retrieved from http://www.qualitative-research.net/index.php/fqs/article/view/1089

McDonald, N. (2010). *Handbook for K-8 arts integration: Purposeful planning across the curriculum.* Boston: Pearson.

Miller, A. (2014). *PBL and STEAM education: A natural fit.* Retrieved from http://www.edutopia.org/blog/pbl-and-steam-natural-fit-andrew-miller

Munster, A. (2005). Why is BioArt not terrorism?: Some critical nodes in the networks of infomatice life. *Culture Machine, 7*(0). Retrieved from http://www.culturemachine.net/index.php/cm/article/view/31

National Art Education Association. (2014, April). *Position statement on STEAM education.* National Art Education Association. Retrieved from https://www.arteducators.org/advocacy/articles/143-position-statement-on-steam-education

Nicolescu, B. (1997, November). *The transdisciplinary evolution of the university: Condition for sustainable development.* Retrieved from http://ciret-transdisciplinarity.org/bulletin/b12c8.php

Ossola, A. (2014, December 3). *Is the U.S. focusing too much on STEM?* Retrieved from http://www.theatlantic.com/education/archive/2014/12/is-the-us-focusing-too-much-on-stem/383353/

Patton, R., & Knochel, A. (2017). Meaningful makers: Stuff, sharing, and connection in STEAM curriculum. *Art Education, 70*(1), 36–43.

Peppler, K. A., & Bender, S. (2013). Maker movement spreads innovation one project at a time. *The Phi Delta Kappan, 95*(3), 22–27.

Pomeroy, S. R. (2012, August 22). *From STEM to STEAM: Science and art go hand-in-hand.* Retrieved from http://blogs.scientificamerican.com/guest-blog/from-stem-to-steam-science-and-the-arts-go-hand-in-hand/

3 Creating a STEAM Map: A Content Analysis of Visual Art Practices in STEAM... 55

Riley, S. (2013, December 18). *Pivot point: At the crossroads of STEM, STEAM and arts integration.* Retrieved from http://www.edutopia.org/blog/pivot-point-stem-steam-arts-integration-susan-riley

Robin, J. (2011). *What project based learning isn't.* Retrieved from http://howtovideos.hightech-high.org/video/265/What+Project+Based+Learning+Isn't

Sanders, M. E. (2008). STEM, STEM education, STEMmania. *The Technology Teacher, 68*(4), 20–26.

Santana, M. (2015). *Dresses' flashing lights aim to attract girls to STEM.* Retrieved from http://www.sun-sentinel.com/features/south-florida-parenting/stages/child/sfp-dresses-flashing-lights-aim-to-attract-girls-to-stem-20150302-story.html

Silverstein, L. B., & Layne, S. (2010). *What is arts integration?* Retrieved from http://artsedge.kennedy-center.org/educators/how-to/arts-integration/what-is-arts-integration

Steele, B. (2013). *Something about STEM drives women out.* Retrieved from http://www.news.cornell.edu/stories/2013/11/something-about-stem-drives-women-out.

STEM to STEAM. (n.d.). Retrieved from http://stemtosteam.org/about/

Stroksdieck, M. (2011). *STEM or STEAM?* Retrieved from http://scienceblogs.com/art_of_science_learning/2011/04/01/stem-or-steam/

Thomas, J. W. (2000). *A review of the research on project-based learning.* San Rafael, CA: Autodesk Foundation. Retrieved from http://www.bobpearlman.org/BestPractices/PBL_Research.pdf

Trilling, B., & Fadel, C. (2009). *21st century skills: Learning for life in our times.* San Francisco: Jossey-Bass.

Vande Zande, R. (2010). Teaching design education for cultural, pedagogical, and economic aims. *Studies in Art Education, 53*(1), 248–261.

Ward, R. A., & Albritton, J. (2015, October). Math meets cubism. *STEAMed,* 23–27.

Watson, A. D. (2015). Design thinking for life. *Art Education, 68*(3), 12–18.

Watson, A. D. (2016). Revving up the STEAM engine. *Art Education, 69*(4), 8–9.

White, H. (2010, November). *STEAM – Not STEM whitepaper.* Retrieved from http://steam-not-stem.com/articles/whitepaper/

Wynn, T., & Harris, J. (2012). Toward a STEM + arts curriculum: Creating the teacher team. *Art Education, 65*(5), 42–47.

Yakman, G. (2008). ST$\sum$@M education: An overview of creating a model of integrative education. In *Research on technology, innovation, design and engineering (TIDE) teaching* (pp. 335–358). Salt Lake City, UT.

Yakman, G. (2012). Recognizing the A in STEM education. *Middle Ground, 16*(1), 15–16.

Yakman, G., & Lee, H. (2012). Exploring the exemplary STEAM education in the U.S. as a practical educational framework for Korea. *Journal of the Korean Association for Science Education, 32*(6), 1072–1086.

Zakaria, F. (2015, March 26). *Why America's obsession with STEM education is dangerous.* Retrieved from https://www.washingtonpost.com/opinions/why-stem-wont-make-us-successful/2015/03/26/5f4604f2-d2a5-11e4-ab77-9646eea6a4c7_story.html

Zylinska, J. (2009). *Bioethics in the age of new media.* Cambridge, MA: MIT Press.

Chapter 4
Design Thinking Gives STEAM to Teaching: A Framework That Breaks Disciplinary Boundaries

Danah Henriksen, Rohit Mehta, and Swati Mehta

> *Let us search ... for an epistemology of practice implicit in the artistic, intuitive processes which practitioners bring to situations of uncertainty, instability, uniqueness, and value conflict.*
>
> ~ Donald Schön

Introduction: A Design Framework for STEAM

In this chapter, we present a multi-threaded argument to suggest how design thinking can be an excellent framework for developing STEAM education. We note that STEAM is broader than mere arts integration in STEM. It reflects a view of education that is more creative, real-world-driven, and problem- or project-based in nature. To develop learning content and experiences that offer creative, authentic, real-world, and problem- or project-driven focus, teachers need more than an argument—they need a guiding framework. We suggest that design and design thinking are natural areas of interconnection with STEAM, both for learners and teachers. These ideas can be used to frame STEAM-based experiences that are more open, creative, project-based, and real-world-driven. Here, we discuss the nature of the connections between design and STEAM and focus on how teachers can use design

---

D. Henriksen (✉)
Arizona State University, Tempe, AZ, USA
e-mail: danah.henriksen@asu.edu

R. Mehta
California State University, Fresno, CA, USA
e-mail: mehta@csufresno.edu

S. Mehta
Michigan State University, East Lansing, MI, USA
e-mail: mehtaswa@msu.edu

© Springer Nature Switzerland AG 2019
M. S. Khine, S. Areepattamannil (eds.), *STEAM Education*,
https://doi.org/10.1007/978-3-030-04003-1_4

thinking practices to help them redesign curriculum to transition from STEM to STEAM.

Integrating the arts and sciences in educational settings is essential, as historical evidence demonstrates that the most effective and innovative STEM practitioners draw on both scientific and artistic knowledge and experience (Piro, 2010; Shlain, 1991; Simonton, 1988). However, in practice, the field of education has struggled to realistically blend these disciplines into a STEAM approach (Jolly, 2014, 2016). In part, this may be because the acronym of STEAM suggests that the approach is merely as simple as plugging art into the STEM fields (Piro, 2010). But we suggest that STEAM, in educational terms, may often be far more intricate than any simple combination of the arts with science, technology, engineering, and/or mathematics. Both the arts and STEM disciplines have long suffered from narrow stereotypes that position the sciences as rigid, analytic, cold, and logical and the arts as softer, more intuitive and emotional, and less logical (Feist, 1998). Yet research and expert practitioner experience often show us that this is not true (Henriksen, 2011; Henriksen & Mishra, 2015; Root-Bernstein & Root-Bernstein, 1999; Simonton, 1988). At times, or in certain contexts, these distinctions may hold. But in many other contexts, the sciences in practice often have elements that are aesthetic, interpretative, intuitive, and cultural, while the arts can also be logical, analytical, rational, and process-driven (Caper, 1996; Snow, 1959).

Disciplinary knowledge and practice varies across contexts, and creative thinking drives much progress and knowledge construction—in the arts, STEM, and other disciplines (Root-Bernstein & Root-Bernstein, 1999). The core of STEAM is about learning that blurs the lines of disciplines and is creative and problem- or project-oriented, with real-world complexity. Thus, STEAM learning, teacher cognition, and teaching practices need to have processes that respect this richness and that help teachers develop creative curricula that is instantiated in real-world learning connections. We propose that design as a discipline and a thinking process is an ideal theoretical framework to tie all these threads together. We argue that design can help teachers think in ways that are more problem-oriented, creative, and real-world in their approaches, thereby blurring the disciplinary boundaries across STEAM. .

The arts and STEM disciplines both function in ways that intersect within, between, and across disciplinary lines (Mishra, Henriksen, and Deep-Play Research Group, 2012). More importantly, real-world discovery and knowledge building in STEM disciplines revolve around skills and contexts—like creativity, problem-solving, and authentic, real-world problems and projects (Connor, Karmokar, & Whittington, 2015)—that are often associated with arts. The heart of STEAM projects is not just about the application of the arts to the sciences or vice-versa. This is not to say that simple combinations of different subject matters like art or STEM might not produce interesting or fruitful approaches to STEAM. But we do assert that STEAM as an educational paradigm is broad and there is immense value in expanding the perspectives on the intersections of arts and STEM that go beyond simple combinations. A simple inclusion of arts in STEM as an additional, occasionally visited lens, may certainly be part of the picture. But as others have recently begun to suggest (Jolly, 2014, 2016; Madden et al., 2013), the heart of STEAM education lies in an interdisciplinary approach, which respects the arts and the sciences alongside other disciplines, by equally engaging the analytical and intuitive, the logical, and the aesthetic. The core of STEAM then is not about just STEM or the arts. STEAM learning is about richly integrating subject matters in transdisciplinary ways that engage people in creativity, problem-solving, and project- or problem-based learning, in issues of real-world impact. This implies moving STEAM into a more inclusive paradigm, beyond the mere connection of art and science, and into an arena that speaks broadly to creative, interdisciplinary, real-world, and inquiry-based learning. Along these lines, STEAM has been defined as such:

> STEAM is an educational approach to learning that uses Science, Technology, Engineering, the Arts and Mathematics as access points for guiding student inquiry, dialogue, and critical thinking. The end results are students who take thoughtful risks, engage in experiential learning, persist in problem-solving, embrace collaboration, and work through the creative process. (Education Closet, n.d.)

Design Melds STEAM Together

While scholars have suggested recent different frameworks for STEAM pedagogy (Kim & Park, 2012), few have considered design thinking as a natural and logical approach to STEAM curriculum design for teachers. By focusing on a theoretical framing that inherently connects the arts and sciences, teachers as well as students can engage in work that integrates disciplines.

Design as a creative and flexible discipline epitomizes the intentional blurring of disciplinary boundaries. It is an interdisciplinary area in which art, science, and other disciplines can intersect around human-centric problems (Buchanan, 2001). However, how researchers perceive the role of design in STEAM varies. Recently, a few scholars noted how design as an art form can function as a useful space for teachers to bring STEAM into their classrooms (Bequette & Bequette, 2012; Peppler, 2013). While this approach may offer STEAM opportunities, it is still limited by the fact that it connects the disciplines at their surfaces, while true integration remains a challenge (Radziwill, Benton, & Moellers, 2015). Instead, it may be helpful to consider design at its deeper interdisciplinary roots as a field and view it as a framework of thinking about STEAM in which artistic and scientific disciplines naturally intersect and in which, the core of STEAM is not just about this intersection but about what it means for learning and inquiry.

In this chapter, we suggest that STEAM involves blurring disciplinary boundaries to frame and solve problems—it involves thinking creatively and working on projects that aim at real-world inquiry. We argue that design thinking provides a framework to streamline this disciplinary integration. In teacher education, we have recently begun to use design thinking as a framework and a way for teachers to reframe their thinking about classroom curricula. While this is significant in student learning contexts, we believe design thinking is also a useful framework for teachers to use as they develop more STEAM-based curriculum. Teachers are the central drivers of the work that students do in the classroom, and a significant problem of practice they often encounter involves lesson design and how to make it more project-based, more creative, and thus for STEM teachers, more STEAM-based.

In this piece, we tackle such problems of practice through a design thinking approach in examples of teachers' processes. We initially discuss how design thinking relates to STEAM, by describing design as a disciplinary crossroads between the arts and sciences and a space for creative problem-solving. We suggest that design thinking provides a framework that STEM (and other) teachers can use in their own thinking and curricular design processes, to construct more creative, engaging, and project-based curriculum. We also present three illustrative case examples of educators who have applied design thinking processes in their own lesson design, as part of their work in a design thinking course in teacher education. In this, they used design thinking as part of their teacher education training, to creatively redesign curriculum to be more creative, more problem- or project-based, and driven by authentic real-world learning. In short, to make lessons more STEAM-based.

In summary, we describe the connections between design thinking as a framework for STEAM more generally and exemplify how STEM educators may themselves work through design thinking to build STEAM curriculum. We begin by providing the theoretical foundations of design as a discipline, arguing for its role as an artistic, scientific discipline for human-centered problem-solving and creativity.

Design and STEAM: Creative, Interdisciplinary, Human-Centered Problem-Solving

Educational policy is often constraining and unsupportive of teacher creativity in lesson or curriculum design (Cohen, McCabe, Michelli, & Pickeral, 2009), particularly for teachers attempting to build integrated or STEM approaches that veer away from textbook curriculum. Teachers, like many people, often feel uncertainty about their own individual creative potential (Cropley, 2016)—making it difficult to identify and enact good solutions in lesson design. Scholars have recently begun to discuss possible approaches toward creative thinking via the path of design thinking. As an interdisciplinary realm, design employs approaches, tools, and thinking skills aimed at helping designers devise more and better ideas toward creative solutions (Kelley & Kelley, 2013). The term "design thinking" refers to cognitive processes of design work (Cross, 2001, 2011; Simon, 1969)—or the thinking skills and practices designers use to create new artifacts or ideas or solve problems in practice. In many ways, the interdisciplinary nature of design, and creative, problem-based approach, makes it a useful framework for STEAM integration—for students certainly but also for teachers. In understanding how design and design thinking can function as an area that connects to STEAM, it is helpful to examine the foundations

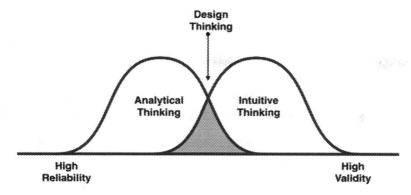

Fig. 4.1 Design thinking

of the discipline. At some level, what we assert here is that design is inherently, in and of itself, STEAM—because it engages the analytical and intuitive both jointly in artistic and scientific ways (see Fig. 4.1).

The arts and sciences are often traditionally spoken of as distinctly different realms which call on different skills. There is a mythology around the idea of the "hard sciences" versus the softer more inimitable artistic disciplines. But creativity researchers and designers both have often pointed out that this is a false dichotomy (Root-Bernstein & Root-Bernstein, 1999). Creativity, as a meta-level thinking skill, is central to STEAM as well as to design, in that it drives most impactful thinking not only in the arts but also in the STEM disciplines. Design highlights the falseness of this dichotomy between arts and STEM, as it naturally fuses them together and acts as a reminder that disciplinary boundaries are blurrier than we often think.

Design lies at the intersection of art and science and applies to a wide range of human-centered disciplines through creative work (Cross, 2011; Weisman, 2012). It is a creative process of intentionally developing something that does not yet exist—something that is novel and effective, and therefore, inherently creative (Cropley, 2001; Fox & Fox, 2000; Oldham & Cummings, 1996; Zhou & George, 2001). Thus, both analytical thinking and divergent creative thinking are keys to design processes (Kelley & Kelley, 2013). A designer's work is iterative and often idiosyncratic, but designers' creativity and design choices are scaffolded and informed by common processes (Buchanan, 2001). These design thinking skills give flexible support and grounding to the open-ended arena of creative practice that lies at the intersection of the arts and sciences (Hoadley & Cox, 2009; Watson, 2015).

Herbert Simon (the Nobel Laureate who founded design as a professional field) offered a definition of design that reflects how applicable it is to human-centered problem-solving:

> Everyone designs who devises courses of action aimed at changing existing situations into preferred ones. The intellectual activity that produces material artifacts is no different fundamentally from the one that prescribes remedies for a sick patient or the one that devises a new sales plan for a company or a social welfare policy for a state. (Simon, 1969 p. 130)

Here, design crosses many fields of human endeavor around complex problems and creative solutions—most notably, constructing knowledge and enacting change. This is evident in the statement that "everyone designs" provided that their goals include "changing existing situations into preferred ones." Buchanan (2001) notes that design involves using human ability for creative problem-solving around ideas, processes, or systems that serve needs. Design involves directing creativity toward goals, actions, and purpose around real-world issues (Collins, Joseph & Bielaczyc, 2004; Hoadley & Cox, 2009). This situates design as a creative problem-solving and thinking approach at the core of human-centered problems and areas, such as teaching, learning, and problem-solving within a STEAM paradigm.

While design has increasingly been noted as a framework for integrating STEAM into what students do, it may also be a productive avenue for teachers to use in their thinking processes as they look toward curriculum design. Using design processes in their own thinking, they may be better able to enact change in the classroom and rethink curricula or lesson design toward more STEAM approaches that are interdisciplinary, creative, and project-oriented. In this, design becomes a way of thinking for teachers as well as learners.

Design Thinking and Teachers: A Path to Creative Learning Design

Many scholars (Kirschner, 2015; Koehler & Mishra, 2005; Norton & Hathaway, 2015) have discussed design as a theoretical lens for teaching and learning. But it has not always been explicitly connected to STEAM, though some have drawn implicit connections. Donald Schön (1983) integrated design as a creative thinking process across disciplines. Schön described how human-centered professions call for "an epistemology of practice implicit in the artistic, intuitive processes which [design and other] practitioners bring to situations of uncertainty, instability, uniqueness and value conflict" (1983, p. 49). This emphasizes design as a creative and reflective action—an ongoing dialogue between processes, people, and materials in real-world problems and work.

Mishra and Koehler (2006) developed their theory of teacher knowledge around the concept of teachers as designers. They underscored the role of educators in working with tools, content, and ideas to design experiences for learners. This suggests that teachers need experiences which place them distinctly in the role of designer, to enhance their knowledge for creative lesson design and crafting learning experiences.

Norton and Hathaway (2015) have noted that teachers are increasingly challenged to be creative in building innovative practices for twenty-first century educational contexts, such as taking STEM to STEAM. Kirschner (2015) describes how the demands of the twenty-first century education, and the creative design aspect of teachers' work, are different from traditional views of teaching as doing or implementing something that already exists. Teachers must be able to be creative designers of learning experiences for students, and this often requires moving traditional STEM work into more STEAM-based learning. Scholars have noted that the professional and creative capacity of teachers is a primary driver and determinant of the experiences of students in school and the types of twenty-first century skills they develop (Darling-Hammond, 2001; Kalantzis & Cope, 2010). But teacher education and professional development has often struggled to give educators specific tools and ways of thinking that help them confront complex and diverse educational problems of practice. For example, a key problem of practice involves creating learning experiences that are more STEAM-based, by way of being more project-based and real-world oriented with a focus on creativity or problem-solving.

As part of our work in teacher education, we developed a course in design thinking, for graduate-level teachers to use in addressing problems of practice. The teachers learn about design thinking and use it to work on and create solutions around their classroom practice. Many of the teachers choose to use design thinking as a framework for helping them to redesign lessons and curricula. Several have aimed at creating lessons that have more elements of a broad STEAM paradigm—in creating curricula that are more creative, project-based, and focused on real-world relevance. In these endeavors, they demonstrate how design thinking is a useful teacher thinking framework, for helping teachers redevelop curricula or lessons toward STEAM. We share several in-depth examples of teachers' design work from the course along these lines. But first, to provide readers more context, we begin with a brief overview of the design thinking model we used: *the Stanford design school model*. Then, we share a brief overview of the course structure and assignments to get a sense of what the teachers were asked to do. Finally, we provide more details through the examples of the teachers' work, followed by synthesizing conclusions and implications about design thinking, STEAM, learning, and teacher education.

Design Thinking in the Stanford Design Model

Design thinking as a term denotes the cognitive processes or thinking skills that designers use to do their work (Watson, 2015). There are many different variations of design thinking models available in the field, most of which have areas of similarity or overlap in themes. And as there is no one best way to approach design thinking, it comes down to exploring and choosing a model that fits well. Design thinking has increasingly been discussed and used to integrate STEAM into more engineering domains, but it also stands by itself as a framework for thinking and problem-solving that spans the arts and sciences. Engineers may use design thinking, but so may visual artists (Boy, 2013; Brophy, Klein, Portsmore, & Rogers, 2008). One of the most popular, commonly noted, and well-established design thinking models is the Stanford design model—created within the Stanford School of Design (Plattner,

2015). This was the guiding model for our teachers to use as they rethought their curriculum in STEAM-based ways, so we provide an overview of it below.

The Stanford model has five phases or stages of design thinking, also referred to as modes, which are worked through toward a problem solution or resolution. These five modes are *empathize, define, ideate, prototype,* and *test*. While we describe them in linear fashion, design thinking is actually an iterative process (Plattner, Meinel, & Leifer, 2010). Designers, teachers, and others can cycle through the process or reenter modes as needed, to understand or explore problems and solutions.

The first mode is **empathize**. Empathy is at the foundation of human-centered design and an essential starting point for any type of design work (Plattner et al., 2010). In this mode designers observe users and their behaviors, interact with and interview them, and try to immerse themselves in understanding the experience and perspective of the user. One might ask questions, listen to stories and experiences, observe their interactions, or explore their world to understand their feelings, ideas, and reasons for behavior. These insights allow designers to approach the rest of the design process with a stronger understanding of the context and problem. Many design models begin the design process with problem identification. The Stanford model requires the designers to first empathize with the people that are experiencing the problem.

In the second mode, the **define** mode, designers use the insights gathered from empathizing to focus in on the problem. They purposefully go beyond a simple definition as they describe the complexities of the user, the problem, and the context. The problem solutions depend on how the problem has been defined. In this mode designers articulate a problem statement based on the understanding they have gained previously. They focus and frame the problem, to guide the design efforts moving forward (Plattner, 2015).

The third mode, **ideate**, explores a wide volume and variety of solutions and ideas (bootcamp bootleg). The goal is to go beyond the obvious to generate far-ranging ideas, solutions, and approaches connected to the problem. Designers must go wide with ideas, keeping the problem in mind but also letting flights of fancy bring up new and creative ideas. Deferring judgment on evaluating ideas provides a sense of freedom and allows for the unconstrained development of ideas.

After designers have generated numerous ideas, they put those ideas into action in the fourth mode of **prototype**, by creating a possible prototype or a model of a solution(s) to the problem (which can later be tested). Prototyping is the act of making ideas concrete. It is not an attempt to arrive at a final solution but an opportunity to jump in and make ideas concrete. A prototype might be a physical object, but it also might be a storyboard, an activity, a drawing, or more.

In the fifth mode of **test**, designers test the prototype with actual or representative users/stakeholders. Designers might interview users, observe them interacting with the prototype, or use any other process to gather feedback for refinement of the solutions. Testing may show that a designer must refine the prototype, or redefine and reexamine the original point of view, or possibly revisit the empathize mode to understand users, or return to the ideate mode to explore alternative solutions.

The framework of these design thinking modes and tools was beneficial in guiding these educators toward new possibilities and solutions. In the next section, we describe the course itself and how the model was applied to the structure of the course and assignments.

Overview of the Course: Design Thinking for Addressing Problems of Practice

"Learning by Design" is an online course offered as part of the Masters of Educational Technology program in a College of Education at a large Midwestern (Big 10) university. The first author of this study was one of the course designers and instructors. The course is fully online, and students were educational professionals from a range of settings and contexts (mostly teachers, with a small mix of administrators, instructional designers, counselors, and others). In this course, design thinking was introduced to be used in ways that serve their own specific and local needs and interests.

The syllabus description states, "this is a course about design. Design as practice and a process. Design as it relates to education and the world around us." The course was broken into seven modules of 2 weeks each, with an introduction module to cover basic ideas about design, followed by a module for each of the phases of the Stanford d-School module (for empathize, define, ideate, prototype, and test) and then a concluding module to finish and summarize. Each module consisted of several key parts, including readings and discussion, a problem of practice project, module labs, and a final reflection paper. A bit more description of each aspect of the work is as follows:

- **Readings and discussion**: This involved ongoing participatory class discourse around design themes, in which individual readings suited to each phase or design topic were assigned, with discussion questions tying these ideas to educational practice and themes. Discussions were both at the whole-class level and in smaller groups.
- **Problem of practice design project**: This was the major semester-long project, in which each student selected a problem of practice area that they approached in their context, and, over the semester, worked on the issue, through each of the phases of the Stanford design model. Each specific phase included associated deliverables, such as an "empathy report" for the empathize phase, to include the findings from their empathy research with the audience/stakeholders, a problem definition statement from the define phase, a record of a brainstorming session (sketches, recordings, images) for the ideate phase, a basic model/mock-up of a possible solution (or solution set) for the prototype phase, and an overview and reflection on the results of a user/audience-test for the test phase.
- **Module labs**: These were shorter, more informal, and creative activities done in each module. Labs were designed to allow students to engage deeply with the phase/theme of the module as an idea, with a focus on creative "out-of-the-box thinking," while engaging with big ideas. Labs were not connected to the larger

design project but simply smaller guided, fun activities. For example, during the empathize phase, students did "A Twice-Told Tale," in which they told a real short story from their own life (or someone they knew), then reimagined, and retold it from another different perspective of someone else involved in the original event/story. They reflected on how a situation might look completely different when you put yourself in someone else's shoes. For the ideate phase, students kept an "Incubation/Idea Journal" with them for the week, in which they informally noted or sketched any interesting or creative ideas that emerged in their thoughts, whenever they had them. They then reflected on the practice of actively notating emergent ideas (as an alternative to schedule brainstorm sessions) for ideation. Each module had an associated lab activity which was geared at more informal and creative approaches to the design thinking phase.

• **Reflection paper**: This was a final paper in which students reflected and looked ahead to their learning and goals around design thinking, with an eye to future practice.

The Stanford model's modes of empathizing, defining, ideation, prototyping, and testing structured core activities that teachers in the course applied as a lens for their educational problems of practice. In the next section, we discuss several examples of teachers who used their design thinking project work for this course to redesign aspects of their students' learning. Through this they reflect on the usefulness of design to help create activities for students that have more STEAM elements. These examples are offered not as "case study" per se in methodological terms but rather as descriptions of illustrative cases that allow readers to better grasp how teachers can work toward design thinking processes that rethink curriculum. This rethinking can allow for more STEAM elements of interdisciplinarity, project-based learning, creativity, and open-endedness.

Building Toward STEAM: Examples in Teachers' Curriculum Development

All teachers in the Learning by Design course identified a problem of practice to focus on and applied design thinking to their problems. The problems of practice that students in the course addressed ranged across many different issues of teaching (from classroom practices, to parent-teacher communication, to varied issues of teaching and learning). As course instructors, we noticed a particular resonance for students who were using design thinking processes to help them rethink curriculum in ways that were more creative and compelling. This rethinking had different aspects based on each teacher's goals and context. However, all have at least some elements of our expanded interpretation of STEAM, as involving learning that is more student-centered, problem- or project-based, creative, real-world, or interdisciplinary. In this section, we take three of our teachers and share their individual cases of taking a design thinking approach to curriculum development to tease apart the intersections of modes of the Stanford design model.

Design Thinking in Biology: Creative Problem-Solving

Adam, the lead biology teacher at his school, redesigned the ninth grade biology curriculum of his school to align with the Next Generation Science Standards (NGSS) by using the Stanford design thinking model to rethink his current practices. His goal was that, while making curriculum align better with NGSS, he would also try to make learning more engaging, more creative, and more project-based— which fits well with the broader vision of STEAM that we and others (Jolly, 2014, 2016) have noted.

Adam perceived design thinking as a process that guides you to design any kind of product you desire to make—which in Adam's case was a more student-centered, engaging, and project- or inquiry-driven STEAM curriculum. In his five-phase design process, Adam employed the first stage of *empathy* by interviewing his 16 ninth-grade students, one general biology teacher, and one advanced placement biology teacher. All the students were chosen at random from those enrolled in the second trimester of Biology B (the second and final freshmen biology class). Few students were chosen from sophomores, juniors, and seniors who had taken the biology class and were interested in providing input. His main aim in this phase was to understand the perceived strengths and weaknesses within the present curriculum.

Adam wanted his students to be more successful in biology, and toward that end, he wanted them to be more excited and engaged in science. During the interviews, Adam asked his students to provide specific feedback regarding the probable changes that they wished to see in the curriculum, the pacing of each unit, the structure of each unit, and the lab/activities used within a given unit. The teachers were also asked to respond to the same questions. Both the teachers and student responses were obtained through a live lesson teaching demonstration followed by classroom discussions. Adam found that while the student feedback desired more lab activities, teachers focused on "specific units or activities" that needed change and a need for "deeper understanding for concepts." In his work, he commented that:

> The design process of empathy allowed me to understand the current curriculum was not as engaging as the teachers thought. The students desired more lab activities, and more work that connected them to what is going on in their world. They wanted more of a long-term project-based approaches, such as what they've done in other classes, and would like to see in biology.

From the interviews, Adam concluded that a constructive redesign for his biology curriculum would require an interconnection across these two sets of student and teacher feedback.

Moving to the *define* phase, Adam used the *5 Whys model* to explore and define his problems of practice. The *5 Whys model* is a design-based interrogative technique that helps people understand the root causes of a problem, by asking "why?" a problem exists multiple times to uncover the reasons behind an issue. For example, one way a teacher approach engages this 5 *Whys* technique after interviewing or observing students to understand why they are not performing might be:

1. Students aren't performing in biology classes. Why?
2. They are not engaged in coursework or class sessions. Why?
3. They find the content boring or unrelated to their world. They don't see the excitement or curiosity that's possible in science. Why?
4. The current curriculum does not reflect any of these things or give them opportunities to see connections or explore. Why?
5. It's based on outdated methods that don't reflect constructive, creative, project-based inquiry and needs to be redesigned with these ideas more in mind. Why?

The number of times a designer asks "why?" might vary, and the answers vary with the information they receive in their context, but the core principle remains the same. We need to understand what students need, and in the case of this biology course, what they needed intersected with STEAM principle and new science standards.

Attempting to define his problems led Aaron to explore the reasons for the curriculum revamp. He realized that the reason for him to consider changing his curriculum was based in the fact that, lately, his students were not performing to the standards that teachers and the school had set for them, because the science instruction in its current form allowed for no in-depth creative, project-based work. Hence, Adam's focus was now to redesign the units with lab activities integrated within them and suggests possible alternative assessments to replace the current exhaustive ones.

Having defined his problem of practice, during the *ideate* phase, Adam held a couple of brainstorming sessions with other science teachers, starting with the following question: "Since the NGSS has been adopted, what are some of the things we need to incorporate into our classrooms to align with the new standards, while making sure we maintain the integrity of our teaching?" In response to this question, teachers' feedback generated some interesting ideas on how to introduce new units with few essential questions, making groups where student choose which of these essential questions they would like to address at the end of each unit to promote more inquiry-based in-depth learning. He did brainstorming around formative assessments with the students and integrated them within each unit, using engaging discussion questions, aligning lessons across disciplines, and, finally, integrating free response-type questions to allow a more blended and interdisciplinary approach in which students could practice writing and thinking more freely about science content. This is a series of subtle but powerful moves, in which a traditional curriculum moves toward more STEAM-based learning.

Building off ideation, Adam moved to the *prototype* phase, which comprised of setting up a test unit implementing the changes that were highlighted through the design thinking process. He started working on new lab activities, redesigning units/lessons to integrate new activities, create collaborative lessons and activities to inte-

4 Design Thinking Gives STEAM to Teaching: A Framework That Breaks...

grate teachers' ideas and prepare essential driving questions and presentations, and design alternate, inclusive, and formative assessments decided in the ideate phase. In this final paper, he discussed his process of thinking in the prototype phase:

> For this prototype phase of the process I created a timeline of activities and lessons in a sample unit plan. I wanted my colleague to have the flexibility to utilize their strengths while keeping the format to the unit intact. The changes I made from our traditional unit to this prototype were not radical at all but more practical and based off the input from the empathize phase. Included in the sample unit plan was the inclusion of a long-term alternate assessment of a presentation based upon essential questions. The essential questions were formatted in a way that students had to complete some research and come up with their reasons why the problem exists and how they could solve it or to predict what would happen if the problem continued to exist.

He added all these pieces into a calendar to visualize the teaching plan, keeping them in synchronization with the overarching expectation of integrating the Next Generation Science Standards (NGSS) curriculum.

During the last phase, the *testing* phase, Adam realized that before revising the entire biology ninth grade curriculum to align with current NGSS standards and more STEAM-based instruction, he should best begin in a focused way, by first aligning one unit to the NGSS curriculum and using it as a test to work on the others. This unit was on genetics. Then, he followed the feedback from the empathy interviews and implemented a restructured curriculum design in two classes. This new curriculum included more open-ended inquiry, as well as science project work that framed a genetics topic around a real-world issue that students had to go in-depth around doing more research into, then discuss what the root causes of the problem might be, and creatively brainstorm on multiple in-roads to the problem. This brought the work more clearly into a STEAM paradigm, in which science problems emerge in real-world scenarios that draw upon multiple disciplines, in complex problems that require us to engage multiple possibilities.

Adam added another teacher to test his prototype unit, to get a second perspective (and some validity) on his revised approach. He designed new open-ended questions to obtain feedback from students and a separate set of questions for the teacher. He randomly picked three students from each of the two classes. In his interviews with them, he found that students who had favored new activities in the empathy interviews were appreciative of the change. Others, who were skeptical before, were open to new activities but expected to take some time to get comfortable. The other teacher, on the other hand, noticed more engagement in his class, deeper questions, and better understanding of content. Getting a second perspective on his prototype gave Adam an unbiased take on his design.

Making Design-STEAM Connections

Overall, according to Adam, the process of design thinking allowed him to reflect on his teaching practices and obtaining multiple perspectives on the curriculum. It helped him evaluate issues that were important to both the students and the teachers alike. An important outcome of this design process was the realization and interest of other teachers in redesigning their own curriculum toward more STEAM-based knowing in the sciences. For Adam, this process of design thinking helped him engage his own creativity around a rethinking of curriculum, which allowed for the creation of work that could engage students' creativity, as he noted:

> The design process has forced me to become a little more open minded about solutions and to allow for radical changes. The process allowed me to consider more radical solutions and that is something that I had not always been comfortable doing. When I am considering solutions to a problem or re-designing something, I am normally laser focused with my solution and do not always consider all of my options. The design model has given me the confidence to know that my first idea does not have to be the final answer, and has be rethinking other aspects of my teaching.

He also underscored that he learned how the process of change is slow but constructive feedback from peers and students can lead to effective results, in bringing curriculum from STEM toward STEAM.

Adam's entire search to refine his curriculum while finding innovative ways of integrating new practices taught him that new was made from existing. Creativity is the process of creating something that is *novel* and *effective* (Fox & Fox, 2000). He learned that existing tools and texts could be rearranged in unique combinations to create something that was unique and, therefore, *novel*. However, it also had to be *effective* for his purposes. The design thinking process helped him test that he found something new and whether it was effective or not—thereby thinking creatively all along. Through design thinking, Adam demonstrated how creative problem-solving could make curriculum redesigning an effective process. While we are careful not to suggest that his initial work here is a perfect example of a complete move to STEAM, it is still a step in the right direction, toward a curriculum redesign that is more inquiry-based, project-based, real-world and creative in its approach. It reveals how design thinking becomes a process for rethinking curriculum that naturally engages the analytic and process-based, and the intuitive or creative.

Breaking Traditions in Spanish: Going to the Real World

Katherine, an elementary Spanish teacher in a Midwestern urban school, used design thinking to help her recreate an aspect of her curriculum in a more interdisciplinary, project-based way, by engaging the social aspect of science in an experience for the fourth grade students in her school. Katherine's example is an interesting one in that her teaching content is not STEM but what she designed took a STEAM-related approach. STEAM approaches are often discussed as involving the infusion of arts into sciences, or at least with a focal point on STEM disciplines. But we forget that STEAM also means that STEM disciplines can be woven into art. Root-Bernstein and Root-Bernstein (1999), among others (Simonton, 1988), have noted that exceptional thinkers across disciplines tend to combine ideas across subject matters and that accomplished thinkers working across the arts and humanities often draw upon ideas from STEM disciplines. The essence of STEAM is to cut across disciplinary boundaries to see how projects and learning unfold in more complex real-world terms.

Within her teaching area of Spanish language, Katherine undertook the issue of teaching her students about water crisis, a major community concern in some Spanish-speaking countries, to help students analyze the importance of clean and safe water usage. The main aim of her project was to introduce her students to new Spanish words using water as the center of conversation. She collaborated with two of her elementary Spanish teacher colleagues to ensure multiple perspectives on redesigning the curriculum based on the needs of students. The interdisciplinary STEAM focus of the project she created led students to not only research the scientific dimensions of water usage but to consider problems that occur across countries, learn about the water cycle, and come up with a way to teach other students about the intersection of these issues.

In her design thinking project, for the *empathy* phase, Katherine focused on understanding students' prior knowledge and experiences on the topic of water crisis. She started by administering a survey comprising five questions to her fourth grade students. An example of one of the survey questions she asked is, "What do you think happens when people don't have clean water?" Following the survey, she randomly selected few of her students to get a better understanding of their daily water usage and their conceptions of the necessity of clean water. Then she introduced her students to a hands-on game where she provided them with clean and dirty water tokens, where dirty water tokens significantly exceeded the clean water tokens in number. This way when they exhausted their clean water tokens, they had to make use of the dirty water tokens. This process helped Katherine gain insight of how her students' experiences of using dirty water shaped their conceptions of water shortage issues at their home and within their own community. She also observed increased sense of empathy among her students, as demonstrated in their acts of sharing clean water tokens with each other to sustain longer.

Having analyzed students conceptual understanding of water issues in the empathy phase, Katherine moved to the *define* phase to explore her problem statement in depth. Like Adam, Katherine used the five-why approach to understand the importance of project-based learning (PBL) in her classroom. Her aim was to understand her students' motivation to engage in project-based learning, their reasons to care about real-world issues in their community, their understanding of the importance of water, and why it is an essential element for human existence. To implement the project-based learning process, Katherine included videos of water usage activities from Spanish-speaking countries to help her students envision theoretical discussions. This was a precursor to helping students integrate both art and science, in designing their own water cycle diagram and labeling at least eight words in Spanish in the diagram. Then the students presented their work to peers in a different section in form of student-created slideshows, posters, or brochures.

Katherine collected data by taking photographs of students' water cycle diagrams, facilitating informal discussions, and recording (audio/video) students' in-class online research activity. Through this process, Katherine observed that students presenting their work to their peers helped them foster meaningful discussions and reflections around community issues in which disciplines connect. Sharing their work helped them gain a sense of ownership, which connected them more to the issue.

Moving to the *ideate* phase of design thinking, Katherine mapped out 6 weeks of project-based classroom activity. Along with her teacher colleagues, she decided to implement formative assessment tests to grade students' progress in their projects. During this phase, she used a journal to record her brainstorming sessions with her colleagues and reflected at the end of each day on the feasibility of implementing these ideas. To keep her students engaged, Katherine also decided to raise funds for a prize for the best presentation and most practically designed water cycle diagram.

In the *prototype* phase, Katherine outlined her observations from the previous phases in a Google document to structure a more organized and shareable conceptualization of her activity. This also helped her to gain a deeper insight into the evolution of her lessons and how she scaffolded her students at each step. This prompted Katherine and her colleagues to realize the complexity of making water issues more understandable and accessible to their students, which resulted in her creating a website for the students: https://sites.google.com/a/apps.harpercreek.net/cleanwaterproject/introduction. The website gave students clear guidelines for the project-based activity and defined the tasks and processes they were required to undertake to complete their projects. Katherine and her colleagues updated this "student-friendly" website with additional resources as an eminent part of her implementation phase, focusing on both student and teacher needs.

The final phase of design, *testing*, helped Katherine reflect on her approach as a two-step process. The first step included conducting interviews and discussions, and observing student knowledge of design and labels in Spanish. The second step consisted of introducing students to different online spaces, providing them with different texts and tools, like games and videos, while she observed their choices and facilitated discussion. Formative assessments allowed her to gain insight into her students' understanding and helped her redefine, modify, and present the problem in different ways to fit the needs of her students. She also ensured that students had time to become familiar with the website, which she used to observe their choices, interact with them, and receive feedback.

Making Design-STEAM Connections

Katherine's real-world approach to problem-based learning provided an opportunity to understand students' individual needs, their interests, and their conceptions of community issues in a STEAM project that blended elements of the arts, humanities, science, and social issues. Her design thinking approach strengthened her bond with her students, which was evidenced in her students' sense of agency in sharing of documentaries and informational videos with each other and her, manifesting their engagement and interest in the activity. For Katherine, this process helped her reflect on her teaching practice, thereby realizing the need for teachers to keep innovating their lessons for student engagement. Reflecting on her traditional curriculum plan, she expressed how design thinking aligned with curriculum development and related to specific components of lesson planning and curriculum development. She credited the cyclical nature of the design process for her enhanced problem-solving skills in building out a STEAM idea into her curriculum and found it to be, in her words, "cohesive to how the brain processes information." She described how she intends to continue this innovative design thinking process throughout different realms in her professional career toward more creative teaching.

Designing Math for Authentic Engagement: Failing Better

Chloe, a teacher in a Midwestern school, focused on redesigning the second-grade math curriculum of her school to incorporate a student-centered, data-driven curriculum that promotes student interest and engagement along multiple lines.

In the *empathize* phase of design thinking, Chloe started by planning to conduct some research with her students. First, she randomly selected a few students at different academic levels and interviewed them to understand their in- and out-of-school experiences, preferences, and motivations. She also obtained some observational data by sitting in math classrooms where students were engaged with worksheets. One of the classrooms she sat in had a technology-centric opportunity for students to "play and learn" with math. Having access to technologically different classrooms gave Chloe a chance to compare and notice that a student who would be considered disruptive in a "traditional" classroom was constructive, productive, and successful in a thoughtfully instantiated, technology-centric class setting. Through these observations and interviews, she realized the value in seeing math as "multidimensional" and interdisciplinary. She started to experiment integrating activities that revolve around, as she described it, "play, manipulatives, edible creations, reciprocal teaching, and technology." This infusion of multiple disciplinary approaches, aiming at engaging creativity and different ways of knowing and learning into the mathematics curriculum, started to move her toward a more STEAM-relevant approach.

Chloe's *define* phase was comprised of *why-how* steps of visualizing the problem. First, the *why's* helped Chloe understand the problem of time constraint. Math teachers in her school wanted to add new coursework to the existing content. Chloe noted that when the existing curriculum had been developed, the approach taken involved using confusing and incomprehensible math binders. Chloe's biggest challenge, therefore, was now to find a balance in the existing content and new, innovative coursework, so she could replace the existing incomprehensible material with exciting, creative, and more project-based material. Her task was to address a group of second-grade students who were excited and eager to work with an innovative math curriculum that challenged them intellectually. Taking a problem-solving approach, she saw potential in seamless integration of technology into the lessons, such as hands-on activities with tools, videos, computer lab time, project-based work, and game-based learning, that help students tie in the concepts taught in the lessons to "out-of-school" experiences. She also saw potential in individualized tablet use as a way of practice for each student. She identified greater problems with assessment when assessing student worksheets in math, since they provided a partial understanding. To make help visualize this problem better, she decided to focus on just one unit and make changes to it according to her plan.

During the *ideate* phase, Chloe brainstormed ideas with colleagues during one of her grade-level team meetings. She approached this by introducing her fellow teachers to her problem of practice and possible solutions, looking for ideas to brainstorm. She asked her peers to think freely on the ideas to replace the lesson worksheets, and she focused on keeping the brainstorming coherent. She jotted down all these ideas on a Stormboard (an online brainstorming and planning space), helping everyone to visualize the brainstormed ideas. This helped her reframe her own ideas as well to better suit the students' needs. This collective reflection process prompted her to look forward to the prototype stage and to refine some of her ideas before adding them to the curriculum.

Following ideation, in the *prototype* phase, Chloe created a structured plan for the new math curriculum that was flexible enough for her peers to use in the future. She aimed at redesigning one unit and prepared activities for 1 week, which focused on first introducing the topic to the students, introducing new learning activities, and through them, on three main components: exploration, collaboration, and evaluation. Each day of her math lessons had a theme that each activity was blanketed under to give students a more coherent experience and a big picture view.

The *test* phase in her redesign included two activities. The first focused on an activity called "Scoot" and the second on taking students to computer labs and using MobyMax to measure their conceptions of math. Scoot involved a set of task cards—with a unique problem on each card, along with a number—that was distributed across the room. Students were asked to walk around at a musical prompt and stop when the music stopped. Then, they had to pick a card closest to them and write down their responses on a record sheet. Throughout this time, they were not supposed to speak to each other nor be on the same task simultaneously. At end of the activity, students discussed their solutions and compared notes. They found this collective problem-solving to be more productive and collaborative. In the MobyMax activity in the computer lab, students went on the MobyMax website to check their mathematical concepts. MobyMax is designed to help identify students' conceptual gaps. After these activities ended, Chloe chose five students to interview in addition to recording her own observations during the activities. Her analysis of the responses and observations yielded a "hundred percent engagement" among students. Her students loved the Scoot activity and preferred it over MobyMax. Students seemed to enjoy MobyMax the most when there were incentives to earn a high score (for example, earning game time or a badge for every five high scores)—but across the board she found success in her endeavor to make mathematics more creative and engaging for students.

Making Design-STEAM Connections

In addition to the success she found in increasing student engagement through more STEAM-based learning, an important takeaway for Chloe was the relevance she saw in looking at failure as a constructive factor. She realized that it was important to allow failure to ensure explorations toward the best solution for the problem at hand—both as a teacher and as a learner. This is significant in that a willingness to fail and see failure as a productive thing has frequently been noted as a part of most creative thinking, work, and processes (Smith & Henriksen, 2016). Chloe also appreciated the significance of empathizing with her students and gaining an in-depth perspective on what is most important to them. She found value in reiteration and repurposing as a productive step in problem-solving. Creating authentic STEAM-related learning experiences for her students involved going through an iterative process of defining problems and then designing a path to finding solutions for those problems. It involved failing several times and then learning from those failures. In process, Chloe learned that her journey of failing again and failing better was the perfect example of what she wanted her students to experience, because learning through failure was what made it authentic and ultimately led to a more creative result.

Conclusions

Across these illustrative mini-cases, we have tried to exemplify and tie together several ideas. The first is to present the process of design thinking (demonstrated here in the specifics of the Stanford design model) as a viable path for teachers to work through in redesigning or rethinking curriculum, to move toward more STEAM-based learning. We suggest that design has a multi-threaded connection to STEAM, both in providing teachers with a process to reconsider curriculum design and also in that design itself intricately weaves between STEM, the arts, and other disciplinary content. In offering examples in action from several teachers in this course, we hope to show how teacher education might consider design as a frame-work for teachers, to blend the analytic with the creative in how they think about curriculum. Through this intricate synthesis of analytic and creative thinking, design is itself a form of STEAM-based learning.

In considering how multiple disciplines intersect in the field of design around human-centered problems, we must also realize that most human-centered problems represent a mixture of disciplines (Buchanan, 2001). And within those problems, disciplinary stereotypes do not always hold true. For example, the sciences can have strong social justice sensibilities when STEM fields come to bear on real-world problems, just as mathematics may have a sense of artistic beauty and awe in a language that explains universal laws. Conversely, the arts can have clean lines, edged precision, skill-based processes, or a sense of straightforward purpose in message. Disciplinary content plays out in a range of ways across fields and contexts—in much richer ways that most conventional bounded subject-matter learning suggests (Root-Bernstein & Root-Bernstein, 1999). Both design and STEAM can work together in ways that respect the messy, interdisciplinary, creative, real-world project-based nature of such teaching.

Without the framework of design, STEAM is sometimes conceived of as a basic integration of arts to STEM—which may have its own advantages and which we do not detract from. But which also does not completely address the full potential of a marriage of multidisciplinary ideas or the blurring of disciplinary lines that practitioners of the arts, STEM disciplines, and all other fields tend to experience in practice.

Design can help us to broaden our view of STEAM as an area of disciplinary intersection. It can also help teachers by offering a framework of design thinking skills that may guide them in the revaluation and redesign of STEM or other curriculum, toward more STEAM. The three cases we have discussed here provide just a small set of examples of the possibilities for ways teachers might consider using design skills in their own processes to support STEAM curricular efforts. When teachers are involved in weaving together STEAM ideas using design thinking, the important themes that emerge and connect these paradigms include several things: a focus on creativity, connections to real-world examples or applications, the use of problem- or project-based teaching and learning, and potential for authentic human-centered experiences. These themes are nothing unheard of or entirely new. They have also emerged often through much of the history of recent work around constructivism and current educational psychology (Sawyer, 2011), in what is known about effective teaching. However, what we are introducing in the consideration of all of these themes taken together is that they provide a connective tissue between the domain of design and STEAM. In this connection, there is much for both research and practice to explore, as we seek to broaden the landscape of STEAM work and the capacity of teachers to infuse it into learning experiences.

References

Bequette, J. W., & Bequette, M. B. (2012). A place for art and design education in the STEM conversation. *Art Education, 65*(2), 40–47.

Boy, G. A. (2013). From STEM to STEAM: Toward a human-centered education, creativity & learning thinking. In *Proceedings of the 31st European conference on cognitive ergonomics* (p. 3). New York: ACM. https://dl.acm.org/citation.cfm?id=2501907&picked=prox

Brophy, S., Klein, S., Portsmore, M., & Rogers, C. (2008). Advancing engineering education in P-12 classrooms. *Journal of Engineering Education, 97*(3), 369–387.

Buchanan, R. (2001). Design and the new rhetoric: Productive arts in the philosophy of culture. *Philosophy and Rhetoric, 34*(3), 183–206.

Caper, R. (1996). Play, experimentation and creativity. *The International Journal of Psycho-Analysis, 77*(5), 859–869.

Cohen, J., McCabe, L., Michelli, N. M., & Pickeral, T. (2009). School climate: Research, policy, practice, and teacher education. *Teachers College Record, 111*(1), 180–213.

Collins, A., Joseph, D., & Bielaczyc, K. (2004). Design research: Theoretical and methodological issues. *The Journal of the Learning Sciences, 13*(1), 15–42.

Connor, A. M., Karmokar, S., & Whittington, C. (2015). From STEM to STEAM: Strategies for enhancing engineering & technology education. *International Journal of Engineering Pedagogies, 5*(2), 37–47.

Cropley, A. J. (2001). *Creativity in education & learning: A guide for teachers and educators*. London: Psychology Press.

Cropley, D. H. (2016). Creativity in engineering. In *Multidisciplinary contributions to the science of creative thinking* (pp. 155–173). Singapore, Singapore: Springer.

Cross, N. (2001). Designerly ways of knowing: Design discipline versus design science. *Design Issues, 17*(3), 49–55.

Cross, N. (2011). *Design thinking: Understanding how designers think and work*. Oxford, UK: Berg.

Darling-Hammond, L. (2001). The challenge of staffing our schools. *Educational Leadership, 58*(8), 12–17.

Education Closet. (n.d.). *What is STEAM?* Retrieved from http://educationcloset.com/steam/what-is-steam/

Feist, G. J. (1998). A meta-analysis of personality in scientific and artistic creativity. *Personality and Social Psychology Review, 2*(4), 290–309.

Fox, J., & Fox, R. (2000). *Exploring the nature of creativity*. Dobuque, IA: Kendall/Hunt Publishers.

Henriksen, D. (2011). *We teach who we are: Creativity and trans-disciplinary thinking in the practices of accomplished teachers*. (Doctoral dissertation). Retrieved from Michigan State University ProQuest Dissertations and Theses.

Henriksen, D., & Mishra, P. (2015). We teach who we are. *Teachers College Record, 117*(7), 1–46.

Hoadley, C., & Cox, C. (2009). What is design knowledge and how do we teach it? In *Educating learning technology designers: Guiding and inspiring creators of innovative educational tools* (pp. 19–35). New York: Routledge.

Jolly, A. (2014). STEM vs. STEAM: Do the arts belong? *Education week: Teacher*. Retrieved from http://www.edweek.org/tm/articles/2014/11/18/ctq-jolly-stem-vs-steam.html

Jolly, A. (2016). *STEM by design: Strategies and activities for grades 4–8*. New York: Routledge.

Kalantzis, M., & Cope, B. (2010). The teacher as designer: Pedagogy in the new media age. *E-Learning and Digital Media, 7*(3), 200–222.

Kelley, T., & Kelley, D. (2013). *Creative confidence: Unleashing the creative potential within us all*. Danvers, MA: Crown Business.

Kim, Y., & Park, N. (2012). Development and application of STEAM teaching model based on the Rube Goldberg's invention. In *Computer science and its applications* (pp. 693–698). Dordrecht, Netherlands: Springer.

Kirschner, P. A. (2015). Do we need teachers as designers of technology enhanced learning? *Instructional Science, 43*(2), 309–322.

Koehler, M. J., & Mishra, P. (2005). Teachers learning technology by design. *Journal of Computing in Teacher Education, 21*(3), 94–102.

Madden, M. E., Baxter, M., Beauchamp, H., Bouchard, K., Habermas, D., Huff, M., et al. (2013). Rethinking STEM education: An interdisciplinary STEAM curriculum. *Procedia Computer Science, 20*, 541–546.

Mishra, P., Henriksen, D., & Deep-Play Research Group. (2012). Rethinking technology & creativity in the 21st century: On being in-disciplined. *TechTrends, 56*(6), 18–21.

Mishra, P., & Koehler, M. J. (2006). Technological pedagogical content knowledge: A framework for teacher knowledge. *Teachers College Record, 108*(6), 1017.

Norton, P., & Hathaway, D. (2015). In search of a teacher education curriculum: Appropriating a design lens to solve problems of practice. *Educational Technology, 55*(6), 3–14.

Oldham, G. R., & Cummings, A. (1996). Employee creativity: Personal and contextual factors at work. *Academy of Management Journal, 39*(3), 607–634.

Peppler, K. A. (2013). STEAM-powered computing education: Using e-textiles to integrate the arts and STEM. *IEEE Computer, 46*(9), 38–43.

Piro, J. (2010). Going from STEM to STEAM: The arts have a role in America's future, too. *Education Week, 29*(24), 28–29.

Plattner, H. (2015). *Bootcamp bootleg.* Institute of Design at Stanford. Retrieved from https://dschool.stanford.edu/wp-content/uploads/2011/03/BootcampBootleg2010v2SLIM.pdf

Plattner, H., Meinel, C., & Leifer, L. (Eds.). (2010). *Design thinking: Understand–improve–apply.* Berlin, Germany: Springer Science & Business Media.

Radziwill, N. M., Benton, M. C., & Moellers, C. (2015). From STEM to STEAM: Reframing what it means to learn. *The STEAM Journal, 2*(1), 3.

Root-Bernstein, R., & Root-Bernstein, M. (1999). *Sparks of genius: The thirteen thinking tools of the world's most creative people.* Boston: Houghton Mifflin.

Sawyer, R. K. (Ed.). (2011). *Structure and improvisation in creative teaching.* Cambridge, UK: Cambridge University Press.

Schön, D. A. (1983). *The reflective practitioner: How professionals think in action.* New York: Basic Books.

Shlain, L. (1991). *Art & physics: Parallel visions in space, time and light.* New York: William Morrow and Company.

Simon, H. A. (1969). *The sciences of the artificial.* Cambridge, MA: MIT Press.

Simonton, D. K. (1988). *Scientific genius: A psychology of science.* New York: Cambridge University Press.

Smith, S., & Henriksen, D. (2016). Fail again, fail better: Embracing failure as a paradigm for creative learning in the arts. *Art Education, 69*(2), 6–11.

Snow, C. P. (1959). *The two cultures and the scientific revolution: The Rede lecture.* New York: Cambridge University Press.

Watson, A. D. (2015). Design thinking for life. *Art Education, 68*(3), 12–18.

Weisman, D. L. (2012). An essay on the art and science of teaching. *The American Economist, 57*(1), 111–125.

Zhou, J., & George, J. (2001). When job dissatisfaction leads to creativity: Encouraging the expression of voice. *Academy of Management Journal, 44*(4), 682–696.

Chapter 5
Investigating the Impact of a Community Makers' Guild Training Program on Elementary and Middle School Educator Perceptions of STEM (Science, Technology, Engineering, and Mathematics)

Jennifer Miller-Ray

Introduction

A lack of research exists examining STEM knowledge base, STEM skill sets, and experiences necessary for teachers to implement STEM-integrated instruction (Nadelson et al., 2013). Stohlmann, Moore, and Roehrig's (2012) explored factors affecting teachers' implementation of a national STEM education program, Project Lead the Way. Research included the following theoretical framework theory employing activities that "build on prior knowledge, organize knowledge around big ideas, include real world situations, foster social discourse, and include a social element" (Stohlmann et al., 2012, p. 30). Instructional activities should include "hands on approaches using manipulative, cooperative learning, discussion, questioning, writing for reflection, problem solving, appropriate integration of technology, and the use of assessment" (Stohlmann et al., 2012, p. 29).

Literature highlights the critical role of teachers' influence in regard to student STEM perceptions. Professional development programs often limit scientific knowledge, and pedagogical experience, and often produce teachers who have limited confidence regarding STEM skill sets (Murphy & Mancini-Samuelson, 2012). Teachers experience a lack of professional development activities focused on improving scientific teaching after completing undergraduate degrees and preservice programs (Cotabish, Dailey, Hughes, & Robinson, 2011).

Nadelson, Seifert, Moll, and Coats (2012) suggested a lack of teachers' exposure to scientific inquiry in postsecondary programs corresponds to a lack of exposure to authentic inquiry models used to validate professional development. Elementary

---

J. Miller-Ray (✉)
Sul Ross State University, Alpine, TX, USA
e-mail: jennifer.miller@sulross.edu

© Springer Nature Switzerland AG 2019
M. S. Khine, S. Areepattamannil (eds.), *STEAM Education*,
https://doi.org/10.1007/978-3-030-04003-1_5

teachers are often the first to introduce students to the STEM pipeline (Nadelson et al., 2012). Unfortunately, research suggests that few elementary teachers engage in professional development to improve scientific instruction (Cotabish et al., 2011). Research that included over 300 primary instructors found strong relationships between scientific professional development and confidence levels in teaching science, suggesting that high quality and sustained professional development is needed (Murphy, Neil. & Beggs, 2007).

Background Understanding Makerspace Learning Environments

Makerspaces, defined as "informal sites for creative production in art, science, and engineering where learners blend digital and physical technologies to explore ideas, learn technical skills, and create new products" offer a new environment to explore STEM concepts (Sheridan et al., 2014, p. 505). Research is needed to further understand how people experience learning in Makerspaces and how this impacts self-efficacy and information behavior (Fourie and Meyer, 2015). Researchers have applied constructionist theories to investigate communication and learning technologies, which build upon designing and creating a tangible artifact of an idea (Sheridan et al., 2014). Constructionist pedagogies encourage teachers to act as a facilitator, while "learning occurs as students' develop new ideas through the making of some type of external artifact. Children become encouraged as they reflect upon and share a personalized representation to gain new knowledge via self-directed learning" (Kafai & Resnick, 1996, pp. 1–2). Constructionism is rooted from constructivism perspectives. Constructionism encompasses the idea that "learning is building a knowledge structure irrespective of the circumstances of learning, but adds to constructivism ideas in that learning happens especially felicitously in a context where the learner is consciously engaged in constructing a public entity" (Papert & Harel, 1991). These spaces are designed to encourage deep engagement with STEM-integrated content, critical thinking, problem-solving, and collaboration while sparking curiosity (Koh & Abbas, 2015). Challenges facing educators interested in providing innovative STEM practice through a classroom Makerspace experience include standardized testing, lack of teacher preparation, and limited access to technology and resources (Hira, Joslyn, & Hynes, 2014).

Knowledge is "being actively constructed by the individual and knowing is an adaptive process within an experiential environment" (Karagiorigi & Symeou, 2005). Constructivism proponents argue that building knowledge occurs inside a learner's head (Stager, 2013; Tangdhanakanond, Pitiyanuwat, & Archwamety, 2006). However, constructionists argue that knowledge transformation occurs as the learner is presented opportunities to build and "make an artifact with their own style" inspiring ownership (Papert & Harel, 1991). Papert (1993) proposed that learners must actively construct something tangible outside of the learner's head, presenting an artifact that is sharable and open to critique, promoting the ability to "show, discuss, examine, and reflect with others on cognitive artifacts and products created" (Tangdhanakanond et al., 2006). The design process "focuses on a meta-representational competence, using tools to support communication of an idea, in which learners problem solve, create a prototype, and assess how it works" (Sheridan et al., 2014, p. 508). As learners have opportunities to make a tangible object of interest, they build new knowledge and reinforce through sharing socially (Tangdhanakanond et al., 2006). Environments facilitating simulations employing exploratory learning enhance problem-solving through an active learning and social context (Li, Cheng, & Liu, 2013).

Current Factors Impacting STEM Education Training Programs

According to the Congressional Research Service Report to Congress (Kuenzi, 2008), there is a confirmed concern regarding STEM preparation programs serving students, teachers, and practitioners. Literature identifies challenges in STEM professional development programs (Nadelson et al., 2013). Teachers do play a critical role in regard to student STEM perceptions. For example, Knezek, Christensen, and Tyler-Wood's (2011) MSOSW (Middle Schoolers Out to Save the World) findings indicated that gaps existed regarding the "perceptions towards science, technology, engineering, and mathematics held by middle school students versus those of their teachers" (p. 111). Findings suggested that the millennial generation's lower perceptions toward STEM and STEM careers versus older generation perceptions could result in a lower standard of living for the millennial generation.

Jang (2016) identified STEM skill sets to include critical thinking, reading comprehension, active listening, speaking, complex problem-solving, judgment and decision-making, writing, monitoring, active learning, time management, coordination, systems analysis, mathematics, social perceptiveness, systems evaluation, instructing, science, and learning strategies. Professional development programs often fail to include a focus on scientific knowledge and pedagogical experiences and may produce teachers who have limited confidence regarding STEM skill sets (Murphy & Mancini-Samuelson, 2012). Few teachers engage in professional development activities to improve scientific teaching after receiving degrees (Cotabish et al., 2011). Without STEM-prepared teachers who have positive dispositions toward STEM, how do we improve middle school student perceptions toward STEM and STEM career pathways?

Federal and education agencies continue to stress the need for teacher professional development programs to integrate technology into the classroom effectively and have promoted improved integration programs for over a decade (Keengwe, Georgina, & Wachira, 2010). Despite having improved access to broadband and expanded infrastructure capabilities, educational technologies have yet to be effectively integrated into most K-12 classroom environments (Keengwe et al., 2010). Teachers lack skill sets and expertise regarding how to use technology and lack pedagogical knowledge in regard to integrating it appropriately (Keengwe et al., 2010). Christensen, Parker, and Knezek's (2005) research suggested that many approaches to integrating technology skills in teacher preparation programs are effective as long as authentic technology integration activities are well designed, participants have access to technology, and instruction is included on the use of technology tools. Additional research has investigated teacher progression through stages to further explore teacher barriers to introducing technologies into K-12 STEM professional development programs (Skaza, Crippen, & Carroll, 2013). Koh and Abbas (2015) highlighted the need for the American Library Association to update curricular competencies to address Makerspace library professionals. Findings suggest a critical need to introduce librarians and Makerspace professionals to approaches that facilitate learning and to improve understanding how to design user-appropriate and hands-on learning (Koh & Abbas, 2015). A challenge facing Makerspace environments is the considerable amount of STEM professional development needed to implement such programs (Hira et al., 2014). Often information professionals and librarians facilitate STEM Makerspace activities, but many lack skills and competencies required to sustain Makerspace programs (Koh & Abbas, 2015).

STEM Professional Development Trends

A new approach to professional development exploring Makerspaces launched by the University of Nevada in 2016 employed a mobile Makerspace (Purpur, Radniecki, Colegrove, & Klenke, 2016). The pop-up mobile Makerspace research outcomes reported an increase in STEM enthusiasm and engagement for experimenting with new forms of technology (Purpur et al., 2016). Participants were exposed to three outreach events, each occurring for around a half an hour, in which participants were introduced to 3D printing, digital design literacies, and lendable technologies (Purpur et al. 2016). STEM professional development research led by the i-STEM summer institute (Nadelson et al., 2012) confirms that community space is an effective component to professional development. This finding is supported by additional research produced by the National Aeronautics and Space Administration (NASA) and California State University System's STEM K-12 professional development's Independent Collaborative Model, which centered on a common theme or NASA mission (Liddicoat, 2008). Professional development should encourage peer coaching, practice, and the ability to experience inquiry-based instruction at a minimum of 45 h annually (Cotabish et al., 2011). Recent research investigated the impact of a 3-day STEM professional development institute on elementary teachers' changes in attitudes, confidence, and self-efficacy (Nadelson et al., 2013). The study found significant evidence indicating that short periods of targeted STEM professional development can greatly influence and improve teacher confidence and self-efficacy (Nadelson et al., 2013).

Technology Integration Training Approaches

How are technologies used to enhance pedagogical knowledge that incorporates constructionism? Alesandrini and Larson (2002) recommend teachers work collaboratively contextualizing, clarifying, inquiring, planning, realizing, testing, modifying, interpreting, reflecting, and celebrating to share artifacts and final accomplishments to a wider audience during professional development. President Obama's Educate to Innovate campaign stresses the importance of creative making experiences in which learning design promotes hands-on activities through informal learning spaces via museums, libraries, and community spaces (Sheridan et al., 2014). Sun et al. (2014) suggest incorporating instructional approaches that merge physical and virtual and offer a design eLearning approach via 3D printing. Digital tools that "develop, challenge, and expand prior thinking to become disrupted can lead to new understandings via a more effective pedagogical approach enabled through new technologies" (Sun et al., 2014, p. 210). Through "rapid prototyping," learners can employ digital fabrication to make anything imaginable, inspiring K-12 creativity, and has shown to positively affect attitudes towards STEM and STEM careers (Smith, 2014).

The TPACK framework supports the use of technology as a support for "content being taught and pedagogical strategies for successful outcomes or confidence" and provides a natural framework toward accessing STEM attitudes and beliefs (Smith, 2014). The Technological Pedagogical Content Knowledge (TPACK) framework "builds on Lee Shulman's (1986, 1987) construct of pedagogical content knowledge (PCK) to include technology knowledge" (Koehler, Mishra, & Cain, 2013, p. 13). Based on Shulman's (1986) theories, Mishra and Koehler (2006) developed an instructional model, TPACK, for twenty-first century learning environments investigating pedagogical knowledge, content knowledge, and technology knowledge (Matherson, Wilson, & Wright, 2014). Literature review concludes TPACK research is still in its infancy, with a need to explore TPACK competencies aligned to content domains, assessment of teacher TPACK competencies, and further development of TPACK instrumentation (Voogt, Knezek, Cox, Knezek, & ten Brummelhuis, 2013).

Digital fabrication technologies are classified into two areas to include 2D technologies in which subtractive techniques are employed to trim materials using paper or metal or 3D technologies that use silicone or plastic material excursions (Smith, 2014). The Smith (2014) case study employed the TPACK framework to address a lack of research exploring pedagogical practices integrating 2D digital fabrication technologies into language arts classrooms. The study did report an increase in motivation through hands-on creation of objects.

How can experiential learning activities enhance STEM skill sets? Research investigating Makerspace environments found that experiential learning activities via digital tools, wood working, electronics, circuitry, design, fabrication, music, art, transportation, and food through a creative space engages all ages, races, and populations and fuels access to just-in-time STEM experiences (Sheridan et al., 2014). Smith's (2014) study investigating experiential learning via 2D digital fabrication provides a digital learning framework in which learners clarify, visualize, prototype, implement, and reflect. Flowers, Raynor, and White (2012) highlight challenges facing STEM online teacher preparation programs and suggest that a wide array of methods for evaluation be incorporated to include student portfolios and STEM-based projects.

Spatial reasoning skill sets are highly desired in STEM careers that require a strong understanding of the relationship between 3D space and objects (Park, Kim, & Sohn, 2011). Spatial visualization tests suggest that spatial visualization skills decrease in levels of performance as learners age and can be improved through training (Park et al., 2011).

Learning can be enhanced through the employment of materials to engage multiple sensory modality (Horowitz & Schultz, 2014). Research suggests that the transfer of learning between 2D and 3D contexts is highly complex, changing gradually during stages of cognitive development and requires careful consideration to best reduce cognitive overload or prevent disruptive learning experiences (Barr, 2010).

Improvements in 2D and 3D technologies have led to more commercially available modeling software and hardware, improved file format conversion processes and portable hardware, and have become relatively inexpensive (Horowitz & Schultz, 2014). Applications to the education setting leads some to consider how rapid 3D prototyping in design education could be leveraged to improve student spatial visualization skill sets (Park et al., 2011). Modeling and 3D printing require supervision along with training, but supervision could be supported through on-demand libraries or outreach centers (Horowitz & Schultz, 2014).

STEM Research Trends

Limited research exists examining STEM knowledge base, STEM skill sets, and experiences necessary for teachers to implement STEM-integrated instruction. STEM professional development research by the i-STEM summer institute confirms that a community Makerspace is an effective component in professional development (Nadelson et al., 2012). This finding appears to be supported by additional research produced by NASA and California State University System's STEM K-12 professional development's Independent Collaborative Model, which centered on a common theme or NASA mission (Liddicoat, 2008). STEM professional development models delivered via STEM outreach were equipped with instructional activities, free science and technology resources, and learning technology equipment could be used to engage and peak teacher interest (Liddicoat, 2008).

Makers' Guild Research Program

This study incorporated the Makers' Guild, a series of STEM and instructional technology professional development activities, over the course of the 2016 spring semester and introduced participants to short and targeted project-based learning activities, or challenge cards, connected to curriculum content to employ in a Makerspace environment. The researcher developed a quantitative study design that investigated the relationship between professional development and teacher's attitudes and confidence levels toward technology integration and attitudes toward STEM. The following research question was explored as part of the Makers' Guild program.

1. To what extent do educators who participate in STEM Makerspace professional development activities increase in their self-appraisal of competence in technology integration?
2. To what extent do educators who participate in STEM Makerspace professional development activities increase in their confidence in integrating new information technology into pedagogical practice as measured by The Technology Proficiency Self-Assessment for 21st Century Learning?
3. To what extent do educators who participate in STEM Makerspace professional development activities become more positive in their attitudes toward STEM?

Research Design and Methodology

The Makers' Guild program targeted 6 schools from a large North Texas public school district encompassing 5 cities and serving over 25,000 students. The Makers' Guild included a sample population of 57 elementary and middle school classroom teachers, campus principals, academic coaches, and librarians. Participants took part in professional development activities over the course of a semester beginning in January 2016 and concluding in May 2016. Additional support will be provided during the summer of 2016, with the expectation that teachers will transfer learning to their classrooms the following year. Learning activities included curriculum content connections to include science, math, and the arts. Teachers were introduced to a series of professional development training experiences in STEAM activities integrating 2D and 3D technologies delivered in face-to-face training opportunities and one online training session.

Course activities integrated programing, drafting programs, digital art, digital media, social media, and creation tools with a library Makerspace program targeting elementary and middle school core content areas. Activities incorporated hands-on constructionist approaches to themes geared to reading programs employed by all core content areas. The researcher partnered with the public library Makerspace community and met at the Makerspot, which served as the primary location for professional development. The public library's Makerspace community, along with district librarians, delivered much of the professional development over the course of 4 months.

The purpose of the Makers' Guild program was to introduce participants to Makerspace environments, Makerspace design, constructionism, project-based learning, connecting Makerspace activities to content areas, and expose participants to 3D technologies, 2D technologies, media arts, virtual learning environments, and STEM. Participating schools were awarded Makerspace equipment through a NASA grant as part of the research study to be designed during professional development activities and open to students during the fall of 2016. Three face-to-face training sessions were held, along with one online training module delivered within Canvas (a Learning Management System), along with site visits to facilitate additional support to each participating school. The online project-based Canvas course facilitated community discussions, provided resources, and will continue to serve as a community repository to exchange Makerspace project-based learning activities.

Participants were introduced to the concept of Makerspace workstations to facilitate STEM career awareness through project-based learning activities. Challenge cards connecting content curriculum to Makerspace environments were introduced. The researcher collaborated with district curriculum and digital learning leaders to create a Makerspace project-based learning process, which was introduced to Makers' Guild participants. The Makerspace project-based literacy process can be seen (Fig. 5.1) below.

The process introduces participants to STEM careers, with dashed lines representing an ability for participants to move freely and experience multiple career paths. Curriculum leaders collaborated with the researcher to develop challenge cards to be placed in one of four stations that connected to curriculum content areas.

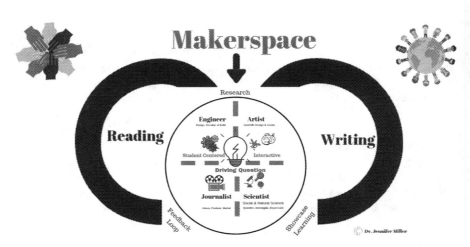

Fig. 5.1 Makers' Guild Project-Based Literacy Process

The challenge cards included a research element to stress the importance of media literacy. Educators were provided challenge card examples in the online course, and challenge cards are planned to be incorporated with students in the fall of 2016. Each challenge card included a research prompt and a short challenge in which learners took on a STEAM career role and challenged to think like an artist, think like a scientist, think like an engineer, or think like a journalist to solve a problem. Examples of challenge cards can be seen below (Figs. 5.2, 5.3, 5.4, 5.5, and 5.6).

School personnel who participated in this research project designed a Makerspace environment to use with students and received green screen equipment, 3D printers, 2D printers, robotic kits, and Makerspace supplies in June 2016. This equipment will be used with students during the fall of 2016 to facilitate workstations that incorporate a variety of Makerspace approaches unique to each campus to include Genius Hour, Makerspace classroom activities, and mobile Makerspace environments. Genius hour is an hour in which students explore a topic for an entire year, research, and make an artifact to share with a wider community. Students reflect on goals and the problem-solving process as part of their genius hour experience. Other approaches will tie Makerspace activities directly to curriculum via challenge cards using the Makerspace project-based learning workstation process. As part of the training experience, participants were introduced to how to create challenge cards and connect Makerspace activities to curriculum content. The workstation model incorporating project-based learning employs a variety of visual art technology tools to include green screen technology, fabrication technology, and robotics. Schools participating in the study were awarded a green screen technology, fabrication technology, or robotics package in the summer of 2016.

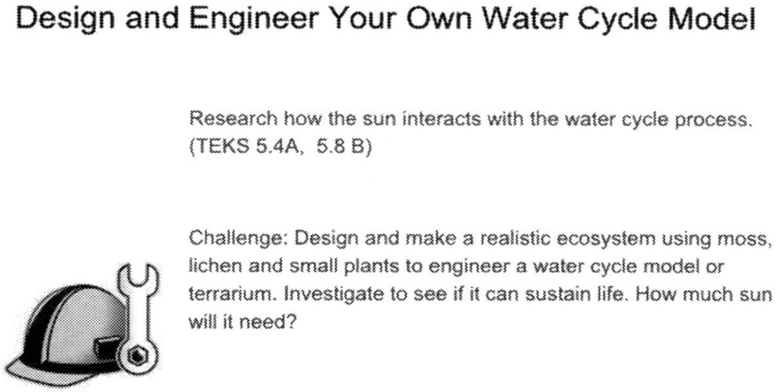

Design and Engineer Your Own Water Cycle Model

Research how the sun interacts with the water cycle process. (TEKS 5.4A, 5.8 B)

Challenge: Design and make a realistic ecosystem using moss, lichen and small plants to engineer a water cycle model or terrarium. Investigate to see if it can sustain life. How much sun will it need?

Fig. 5.2 Engineer challenge card

Water Cycle Art Challenge

Research: Research how the water processes of precipitation, evaporation, and condensation How do they connect to weather? (TEKS 2.8C).

Challenge: Create a work of art that illustrates the water cycle process and how it connects to weather.

Fig. 5.3 Artist challenge card

Hydroelectric Power Investigative Report

Research information about urbanization and water quality. How does urbanization affect the water cycle? How can water be utilized as an alternative water source? (TEKS 5.7 C)

Challenge: Create a public service announcement or blog on the effects of urbanization on water resources and the uses of hydroelectric power.

Fig. 5.4 Journalist challenge card

Instrumentation

A review of literature identified appropriate instruments along with fiscal feasibility of instrumentation appropriate to the proposed study. Three instruments previously used in similar studies were selected to improve internal reliability and validity of the study. The STEM Semantics Survey or SSS (Tyler-Wood, Knezek, & Christensen, 2010) was selected as it was successfully used to measure teacher and student attitudes toward STEM in the Middle Schoolers Save the World (MSOSW) program, which is part of National Science Foundation's Innovative Technology Experiences for Students and Teachers (ITEST). The STEM Semantics Survey is a result of previous modifications from Knezek and Christensen's (1998) Teachers' Attitudes Towards Information Technology questionnaire (TAT), which employed "Semantic differential adjective pairs derived from Osgood's evaluation dimension" (Knezek et al., 2011, p. 94). Targeted statements exploring five scales to include science,

Natural Scientist Investigates Water in Earth's Hydrosphere

Research the meaning of hydrosphere and why it is important for the scientist to study the condition of the surface waters. How do scientists measure the condition of surface waters? (TEKS 5.4A)

Challenge: Make a prediction regarding your local hydrosphere today. Measure and record your local water temperature and water clarity. Explain if your prediction was accurate. Did you understand your data?

Fig. 5.5 Scientist challenge card

Mapping a Watershed History

Research maps and/or area photographs of Birdville ISD and the surrounding areas. Identify different physical features, landforms and water features to include defining the boundaries of local watershed. What changes have occurred? How have people adapted over time? (Social Studies TEKS 5.6.A, B)

Challenge: Make a timeline video or display depicting physical features to include watershed changes over time. Show how people have adapted over time.

Fig. 5.6 Social scientist challenge card

math, engineering, technology, and STEM careers are provided to participants along with seven choices. Internal consistency reliability ratings for all scales are in the range of "very good to excellent," according to DeVellis' (1991) standards, ranging from 0.78 to 0.94 across five constructs for baseline data (Knezek et al., 2011).

An updated version of the TPSA, the Technology Proficiency Self-Assessment for 21st Century Learning (TPSA C-21), was recently updated and expanded to explore twenty-first century learning technologies (Christensen & Knezek, 2014). The new TPSA C-21 encompasses six items to include email, World Wide Web (WWW), emerging tools, integrated applications, teaching with technology, and teaching with emerging technologies (Christensen & Knezek, 2017). The TPSA C-21 was employed to measure the effect of professional development on teachers' attitudes and confidence toward technology integration participating in the Makers' Guild. This research employed a version of the Technology Proficiency Self-Assessment for 21st Century Learning that measures six factors: (F1) email, (F2) World Wide Web, (F3) integrated applications, (F4) teaching with technology, (F5) emerging technologies for student learning, and (F6) emerging technologies for teacher professional development. However, the fourth factor, teaching with technology, produced a low reliability estimate because the version administered included only two of the five items normally used for Factor 4. Internal consistency reliabilities for the six scales ranged from 0.954 to 0.592, considered "very good to poor" according to guidelines provided by DeVellis (1991) with 0.592 representing Factor 4.

Finally, the Stages of Adoption of Technology instrument (Christensen, 1997) was used to investigate the level of teachers' attitudes toward teaching with technology over a period of time. The Stages of Adoption was adapted from Russell's (1995) research exploring how adults utilized new technologies over a period of time and categorizes six stages: (a) awareness, (b) learning the process, (c) understanding the application of the process, (d) familiarity and confidence, (e) adaptation to other contexts, and (f) creative applications to new contexts.

Results

Educators were administered the Stages of Adoption questionnaire, which placed each in one of six stages, prior to receiving training in January 2016 and again at the conclusion of training in April 2016. Out of the 52 subjects who completed both the pre- and posttest Stages of Adoption of Technology survey, 12 moved up at least one category, 33 stayed the same, and 6 moved down at least one category. Twenty participants marked the highest category when completing the pretest Stages of Adoption questionnaire, selecting the "Creative Applications to New Contexts" stage.

The mean scores, standard deviations, and number of all participants are reflected in Figs. 5.7 and 5.8, with the January pretest administration mean of 5.25 and the posttest administration mean at 5.48. Hypothesis 1 was tested using a paired t-test comparing pretest to posttest Stages of Adoption questionnaire score. No significant differences ($p < 0.05$) were found. An analysis of variance (ANOVA) for gender found no statistically ($p < 0.05$) significant difference between male and female responses. Cohen's d for pre- to postscores yielded a small effect with the change in stages of adoption pre- to posttest results not found to be significant ($p < 0.05$). Results depict an increase in the mean from pre to post for all respondents. A one-way analysis of variance (ANOVA) indicated no significant differences ($p < 0.05$) with regard to educators' stage of adoption based on campus socioeconomic status. An increase in attitudes was noted for female teachers, with an effect size of 0.338 pre to post indicating a small to moderate effect (Cohen, 1988) and educationally meaningful according to commonly accepted guidelines (Bialo and Sivin-Kachala, 1996). However, the p levels were not found to be significant ($p < 0.05$). Therefore, the researcher could not conclude the gain was not due to chance. The overall trend indicates that female teachers improved pre to post but that it could have been due to chance. Results do illustrate that leaders ($N = 11$) reported a higher level of competence in technology integration during the pretest administration, which was found to be statistically significant compared to teachers ($p < 0.05$).

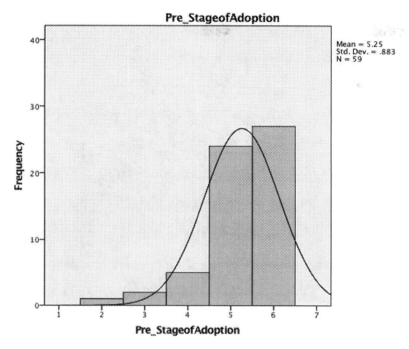

Fig. 5.7 Stages of adoption January pretest questionnaire results for educators participating in Makers' Guild professional development activities

Results indicate that professional development presented through Makerspace environments may increase educators' confidence levels toward integrating technology, especially teachers serving low-income students. Ashbrook (2013) highlights the importance of planning activities for learners to work on a problem or challenge, which promotes STEM inquiry. One way to connect early interest in and the pursuit of STEM careers includes project-based learning activities connected that are applicable to the real world (Christensen & Knezek, 2015a, 2015b, 2017). Activities presented to Makers' Guild participants incorporating project-based learning challenge cards through four STEAM career workstations may encourage an increase in attitudes toward math, science, technology, and STEM careers.

A paired sample t-test was administered to this set of participant data. Table 5.1 depicts the Technology Proficiency Self-Assessment for 21st Century Learning January pretest and April posttest means, number of responses, and standard deviations. Results indicate a positive group mean increase in all factors investigated, suggesting positive improvement in regard to educators' confidence levels. The likelihood of all six measures exhibiting positive changes from pre to post simply by chance would be $p = 0.0156$ using GraphPad Prism version 6.00 for Windows, GraphPad Software, La Jolla California USA, www.graphpad.com.

No significant ($p < 0.05$) pre to post gains were identified for four out of the six factors: (F2) World Wide Web, (F3) integrated applications, (F4) teaching with tech-

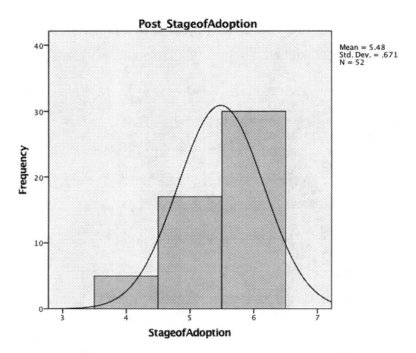

Fig. 5.8 Stages of adoption April posttest questionnaire results for educators participating in Makers' Guild professional development activities

Table 5.1 Descriptive statistics for TPSA C-21 pre-post scores for all respondents participating in Makers' Guild professional development activities

Pre-post scales	Mean	N	Standard deviation	Sig.	Effect size
TPSA email pretest	4.82	52	0.26		
TPSA email posttest	4.90	52	0.22	0.030	0.317
TPSA WWW pretest	4.67	52	0.39		
TPSA WWW posttest	4.74	52	0.28	0.106	0.199
TPSA integrated app pretest	4.43	52	0.65		
TPSA integrated app posttest	4.53	52	0.66	0.109	0.165
TPSA teaching with technology pretest	4.36	52	0.68		
TPSA teaching with technology posttest	4.50	52	0.71	0.084	0.205
TPSA student learning pretest	4.13	51	0.99		
TPSA student learning posttest	4.45	51	0.64	0.007	0.385
TPSA teacher PD pretest	4.68	51	0.44		
TPSA teacher PD posttest	4.79	51	0.36	0.053	0.262

5 Investigating the Impact of a Community Makers' Guild Training Program… 93

nology, and (F6) emerging technologies for teacher professional development. However Factor 1, email skills; Factor 5, emerging technologies for student learning; and Factor 6, teacher professional development were found to have exhibited statistically ($p < 0.05$) significant gains. Educators were more confident in their technology proficiencies in the areas of email skills and using emerging technologies for student learning at the end of Makers' Guild professional development activities than at the beginning.

For additional analyses, participants were categorized into two groups – classroom teachers and school leaders. Teachers included trained and certified classroom professionals working in a classroom serving students from 5 to 13 years of age. Leaders were defined as a certified principal either serving the role of an instructional coach or campus principal. Analysis of variance confirmed significant ($p < 0.05$) differences based on teacher or leader occupation for three of six TPSA C-21 scales at the time of the pretest survey administration: (F3) integrated applications, (F5) emerging technologies for student learning, and (F6) emerging technologies for teacher professional development. The self-appraisal by educational leaders was higher than for teachers for all three scales (Table 5.2). Leaders were more

Table 5.2 TPSA C-21 pretest AVOVA comparing two educator occupations participating in Makers' Guild professional development activities

Sum of squares			df	Mean square	F	Sig.	Effect size
TPSA email	Between groups	0.118	1	0.118	1.087	0.301	0.200
	Within groups	6.189	57	0.109			
	Total	6.307	58				
TPSA WWW	Between groups	0.247	1	0.247	0.972	0.328	0.194
	Within groups	14.506	57	0.254			
	Total	14.754	58				
TPSA integrated app	Between groups	2.774	1	2.774	5.692	0.020	0.446
	Within groups	27.778	57	0.487			
Total		30.551	58				
TPSA teaching with tech	Between groups	1.640	1	1.640	2.563	0.115	0.283
	Within groups	36.462	57	0.640			
	Total	38.102	58				
TPSA student learning	Between groups	4.412	1	4.412	4.137	0.047	0.632
	Within groups	60.794	57	1.067			
	Total	65.207	58				
TPSA teacher PD	Between groups	1.280	1	1.280	4.756	0.033	0.419
	Within groups	15.336	57	0.269			
	Total	16.616	58				

Table 5.3 TPSA C-21 posttest ANOVAs comparing two educator occupations participating in Makers' Guild professional development activities

		Sum of squares	df	Mean square	F	Sig.	Effect size
TPSA email posttest	Between groups	0.001	1	0.001	0.023	0.880	0.044
	Within groups	2.519	50	0.050			
	Total	2.520	51				
TPSA WWW posttest	Between groups	0.002	1	0.002	0.030	0.863	−0.054
	Within groups	3.983	50	0.080			
	Total	3.985	51				
TPSA integrated app	Between groups	0.983	1	0.983	2.314	0.135	0.606
Posttest	Within groups	21.235	50	0.425			
	Total	22.218	51				
TPSA teaching with tech posttest	Between groups	0.461	1	0.461	0.903	0.347	0.334
	Within groups	25.539	50	0.511			
	Total	26.000	51				
TPSA student learning posttest	Between groups	0.198	1	0.198	0.478	0.492	0.250
	Within groups	20.330	49	0.415			
	Total	20.529	50				
	Between groups	0.027	1	0.027	0.204	0.654	0.081

confident entering training than teachers in the technology proficiencies of integrated applications, emerging technologies for student learning, and emerging technologies for teacher professional development at the beginning of the Makers' Guild professional development program. As shown in Table 5.3, there were no significant ($p < 0.05$) differences with regard to occupation at posttest administration on any of the six scales. Based on the group mean averages in regard to occupation, it appears that the self-appraisals of teachers and leaders with respect to their confidence in technology proficiencies became more closely aligned by the end of the professional development activities.

Statistically significant ($p < 0.05$) increases in confidence levels toward emerging technologies for student learning and World Wide Web skills emerged. Leaders reported a statistically significant ($p < 0.05$) higher confidence level in integrated applications, emerging technologies for student learning, and emerging technologies for teacher professional development compared to teachers during pretest activities.

5 Investigating the Impact of a Community Makers' Guild Training Program…

Table 5.4 Paired samples T-test results for STEM Semantics Survey scales for all respondents participating in Makers' Guild professional development activities

		Mean	Standard deviation	T	df	Sig. (1-tailed)	Effect size
Science	Pretest-posttest	0.25000	0.78403	−2.299	51	0.013	0.322
Math	Pretest-posttest	0.83462	1.68998	−3.561	51	0.0005	0.545
Engineering	Pretest-posttest	0.26538	1.12840	−1.696	51	0.048	0.252
Tech	Pretest-posttest	0.40385	1.12476	−2.589	51	0.0065	0.430
STEM career	Pretest-posttest	0.40769	1.43457	−2.049	51	0.023	0.339

Participants were administered the STEM Semantics Survey prior to training in January 2016 and at the conclusion of training in April 2016. Out of 52 participants who completed both the pretest and posttest STEM Semantics Survey questionnaire, all reported an increase in perceptions toward STEM. Internal consistency reliabilities for the six scales ranged from 0.939 to 0.788, considered "excellent to good" according to guidelines provided by DeVellis (1991).

The researcher conducted a paired samples t-test comparing pretest and posttest survey administration scale scores. Of the five areas accessed, there were positive and statistically significant ($p < 0.05$) increases in STEM perceptions for science, math, technology, and STEM as a career. Surprisingly, participants reported the strongest positive increase in perceptions toward Math, with a p value at 0.001, as illustrated in Table 5.4. Effect sizes indicating the magnitude of the gain in each area assessed were (from smallest to largest): 0.252 for perceptions toward engineering, 0.322 for perceptions toward science, 0.339 for perceptions toward STEM as a career, 0.430 for perceptions toward technology, and 0.545 for perceptions toward math. Pre to post gains range from a small effect (0.2 standard deviations) (Cohen, 1988) to a moderate effect (0.5 standard deviations) (Cohen, 1988). The four STEM disposition measures that exhibited statistically significant ($p < 0.05$) gains all are in the range that would be considered educationally meaningful according to commonly accepted guidelines (Bialo & Sivin-Kachala, 1996), and all lie within the zone of desired effects as outlined by Hattie (2009). These analyses confirmed that Makerspace Guild educators did become more positive in their perceptions of math, science, technology, and STEM as a career between the start and the end of professional development.

Discussion and Conclusions

It was observed that the cohort Makers' Guild fostered a sense of community. Teachers seemed to be more excited and inclined to try new technologies because leaders participated in the professional development program, providing value to the school cohort group. The campus cohort groups were observed to be an asset as educators represented a variety of content backgrounds and conversations on connecting Makerspace activities to content areas naturally developed. Educators did report an increase in attitudes toward technology integration. It is difficult to report whether participants would have experienced the same positive trajectory if they have participated in former Makerspace professional development. Activities were presented by the Makerspace community, modeling an active Makerspace community to participants. It was observed that community connections and extended partnerships provided through the public library's Makerspace community strengthened relationships between participating schools and community stakeholders.

It has been observed that teachers' confidence in one's competence in technology integration as measured by the TPSA is an important contributor to success in the classroom (Chrisentsen & Knezek, 2014). Research produced a statistically significant increase in educators' confidence levels in integrating new information technology into pedagogical practice during Makers' Guild professional development activities. Activities were designed to engage participants in an established Makerspace environment. During the first training, educators were slow to participate in Makerspace activities, and many began the training session observing workstations and the Makerspace community. When challenged with the freedom to make any artifact, most teachers did not know how to respond as they seemed to want structure. Most participants had never seen a 3D printer or built a robot, but the Makerspace community was proactive at encouraging participants to try new technologies and experiment with new creative approaches. The social aspect of the community encouraged educators to make an artifact and increased their confidence levels toward integrating emerging technologies for student experiences.

After the first training, resources, communication, and further reading on how Makerspace activities could connect with curriculum were communicated by the researcher and Makers' Guild through emails and Canvas announcements. It was through this platform that participants began to consider project-based learning activities. Challenge cards with curriculum examples were provided to participants, and an example is provided below.

It was observed that content teachers and leaders were very interested in connecting Makerspace activities to curriculum content. The Learning in 3D workshop modeled how this concept could be connected to curriculum with all activities centering around math, science, and vocabulary. Participants were exposed to new applications framed around a curriculum standard. It was observed that online support along with challenge card activities increased participant confidence levels toward the World Wide Web and Emerging Technologies for Student Learning. Teacher confidence levels toward Emerging Technologies for Student Learning and Teacher PD increased because learning experiences were active in nature, with participants making artifacts and sharing with a wider community, particularly for female teachers and teachers serving low-income students. This finding may suggest that further research is needed to explore how the Makerspace environment may contribute to increasing female teacher confidence levels and teachers serving low-income student populations.

Many of the augmented reality and 3D modeling applications introduced to teachers were web based. It was observed that low-income teachers were excited to try challenge cards with students to improve academic vocabulary. Many of the applications and examples used during training incorporated web-based applications in which participants would create an artifact to meet a mathematical or scientific challenge. Many of the challenges emphasized vocabulary activities, and all of the participating schools identified vocabulary as a continuous improvement goal. In addition, it was observed that the online project-based learning Canvas course impacted the confidence levels of integrating the World Wide Web to participants serving low-income student populations. This might explain why educators serving low-income students reported a higher confidence level integrating World Wide Web on posttest results from low-income campuses. Also, an increase in confidence levels toward the World Wide Web could be due to the blended learning grant initiatives. The community exchange offered in the online professional development course was an entirely new experience for all participants in the program.

Christensen and Knezek (2017) stress the importance of STEM proficiency and interest in STEM in elementary and middle school as they have a large impact in students' academic performance and interest in entering a STEM career pathway. Teacher quality in regard to knowledge of the subject matter is "now understood as the greater predictor of academic success," and teachers have little to no STEM training experiences (Liddicoat, 2008, p. 14).

Research did produce a statistically significant finding toward increasing educators' attitudes toward STEM. Many participants in the Makers' Guild had little to no STEM training experience and lacked insight on how STEM could integrate to content areas prior to training. The Makerspace community offered educators the opportunity to see how integrated STEM activities could engage students in a variety of content areas. Activities introduced to educators included a strong math and science connection. For example, educators were introduced to scientific augmented reality interactive word walls, which could be used to improve scientific vocabulary. Scaling methods incorporating 3D design and fabrication printing provided strong connections to math content areas. Measurement conversion activities and story writing introduced through robotics connected both English language arts and mathematics content areas.

Conversations erupted on how such activities could extend classroom content through a creative space for students. Teachers and leaders began to recognize that Makerspace activities could be approached as an extension to curriculum content. Purposeful design could provide a level of engagement for students to consider curriculum content in a Makerspace environment to take on the role of a STEM career professional, serving the role of a scientist, engineer, artist, or journalist. Site visits later emphasized this level of excitement as campuses began to design their Makerspace environment to facilitate STEM career workstations incorporating fabrication printing, robotics, and green screen technologies. Activities incorporated many visualization activities surrounding topics in math and science. Further research exploring visualization technologies, purposeful design, and Makerspace design is needed.

Overall the Makers' Guild professional development experience appears to have been a success. Educators' confidence levels regarding technology and attitudes toward technology and STEM, especially for female teachers and teachers serving low-income populations, did increase. Future research is needed as this study was limited to a treatment group study. A future comparison study could further explore the impact of the Makerspace environment. In addition, future studies are needed to investigate female teacher confidence levels toward technology and attitudes toward STEM and technology in a Makerspace professional development program. Activities incorporated the arts and visualization technologies, with participants creating artifacts using augmented reality, 3D modeling, and origami. Perhaps these activities influenced the increase in teacher confidence levels toward technology and perceptions toward STEM. Future research is needed to explore the art component's impact using the project-based learning process on both students' and educators' perceptions toward math and science in future studies.

References

Alesandrini, K., & Larson, L. (2002). Teachers bridge to constructivism. *The Clearing House, 75*(3), 118–121.

Ashbrook, P. (2013). The STEM of inquiry. *Science and Children, 51*(2), 30–31.

Barr, R. (2010). Transfer of learning between 2D and 3D sources during infancy: Informing theory and practice. *Developmental Review, 30*(2), 128–154.

Bialo, E. R., & Sivin-Kachala, J. (1996). The effectivenss of technology in schools: A summary of recent research. *School Library Media Quarterly, 25*(1), 51–57.

Christensen, R. (1997). Effect of technology integration education on the attitudes of teachers and their students. Unpublished doctoral dissertation, University of North Texas, Denton, TX. Retrieved March 17, 2016, from http://digital.library.unt.edu/explore/collections/UNTETD/

Christensen, R., & Knezek, G. (2014). The technology proficiency self-assessment questionnaire (TPSA): Evolution of a self-efficacy measure for technology integration. In T. Brinda, N. Reynolds, R. Romeike (Eds.), *Proceedings of KEYCIT 2014 – Key competencies in informatics and ICT* (pp. 190–196).

Christensen, R., & Knezek, G. (2015a). The technology proficiency self-assessment 1uestionnaire (TPSA-C21): Evolution of a self-efficacy measure for technology integration. In T. Brinda, N. Reynolds, R. Romeike, & A. Schwill (Eds.), *Proceedings of the KEYCIT 2014: Key competencies in informatics and ICT conference* (p. 311). Potsdam, Germany: University of Potsdam.

Christensen, R., & Knezek, G. (2015b). Active learning approaches to integrating technology into middle school science classrooms: Reconceptualizing a middle school science curriculum based on 21st century skills. In X. Ge, D. Ifenthaler, & J. M. Spector (Eds.), *Full Steam ahead: Emerging technologies for STEAM*. New York: Springer Academic.

Christensen, R., & Knezek, G. (2017). Validating the technology proficiency self assessment for 21st century learning (TPSA C21) instrument. *Journal of Digital Learning in Teacher Education*. https://doi.org/10.1080/21532974.2016.1242391

Christensen, R., Parker, D., & Knezek, G. (2005). *Advances in preservice educator competence and confidence in technology integration: Comparative findings from two Pt3 projects. Integrated technologies, innovative learning: Insights from the PT3 program*. Eugene, OR: ISTE.

Cohen, J. (1988). *Statistical power analysis for the behavioral sciences* (2nd ed.). Hillsdale, NJ: Lawrence Earlbaum Associates.

Cotabish, A., Dailey, D., Hughes, G. D., & Robinson, A. (2011). The effects of a STEM professional development intervention on elementary teachers' science process skills. *Research in the Schools, 18*(2), 16–25.

DeVellis, R. F. (1991). Guidelines in scale development. In *Scale development: Theory and applications* (p. 5191). Newbury Park, CA: Sage.

Flowers, L. O., Raynor, J. E., & White, E. N. (2012). Evaluation in online STEM courses. *International Journal of Business and Social Science, 3*(18), 16–20.

Fourie, I., & Meyer, A. (2015). What to make of makerspaces: Tools and DIY only or is there an interconnected information resources space? *Library Hi Tech, 33*(4), 519–525.

Hattie, J. A. C. (2009). *Visible learning: A synthesis of 800+ meta-analyses on achievement.* Abingdon, UK: Routledge.

Hira, A., Joslyn, C. H., Hynes, M. M. (2014). Classroom Makerspaces: Identifying the opportunities and challenges. In *Proceedings of the 2014 IEEE frontiers in education conference (FIE)* (pp. 1–5). New York: Institute of Electrical and Electronics Engineers (IEEE).

Horowitz, S. S., & Schultz, P. H. (2014). Printing space: Using 3D printing of digital terrain models in geosciences education and research. *Journal of Geoscience Education, 62*(1), 138–145.

Kafai, Y. B., & Resnick, M. (Eds.). (1996). *Constructionism in practice: Designing, thinking, and learning in a digital world* (pp. 1–2). London: Routledge.

Karagiorgi, Y., & Symeou, L. (2005). Translating constructivism into instructional design: Potential and limitations. *Journal of Educational Technology & Society, 8*(1), 17–27.

Keengwe, J., Georgina, D., & Wachira, P. (2010). Faculty training strategies to enhance pedagogy-technology integration. *International Journal of Information and Communication Technology Education (IJICTE), 6*(3), 1–10.

Knezek, G., & Christensen, R. (1998). Internal consistency reliability for the teachers attitudes toward information technology (TAT) questionnaire. In S. McNeil, J. D. Price, S. Boger-Mehall, B. Robin, & J. Willis (Eds.), *Proceedings of the society of information technology & teacher education (SITE)'s 9th international conference* (Vol. 2, pp. 831–832). Charlottesville, VA: Association for the Advancement of Computing in Education.

Knezek, G., Christensen, R., & Tyler-Wood, T. (2011). Contrasting perceptions of STEM content and careers. *Contemporary Issues in Technology and Teacher Education, 11*(1), 92–117.

Koehler, M. J., Mishra, P., & Cain, W. (2013). What is technological pedagogical content knowledge (TPACK)? *Journal of Education, 193*(3), 13–19.

Koh, K., & Abbas, J. (2015). Competencies for information professionals in learning labs and makerspaces. *Journal of Education for Library and Information Science, 56*(2), 114.

Kuenzi, J. J. (2008). *Science, technology, engineering, and mathematics (STEM) education: Background, federal policy, and legislative action.* Washington, DC: Congressional Research Service, Library of Congress.

Li, Z. Z., Cheng, Y. B., & Liu, C. C. (2013). A constructionism framework for designing game-like learning systems: Its effect on different learners. *British Journal of Educational Technology, 44*(2), 208–224.

Liddicoat, S. (2008). NASA enriched collaborative STEM K-12 teacher professional development institutes within the California State University system. In *2008 IEEE frontiers in education conference (FIE)* (pp. 14–19). New York: Institute of Electrical and Electronics Engineers (IEEE).

Matherson, L. H., Wilson, E. K., & Wright, V. H. (2014). Need TPACK? Embrace sustained professional development. *Delta Kappa Gamma Bulletin, 81*(1), 45–52.

Mishra, P., & Koehler, M. (2006). Technological pedagogical content knowledge: A framework for teacher knowledge. *The Teachers College Record, 108*(6), 1017–1054.

Murphy, C., Neil, P., & Beggs, J. (2007). Primary science teacher confidence revisited: Ten years on. *Educational Research, 49*(4), 415–430.

Murphy, T. P., & Mancini-Samuelson, G. (2012). Graduating STEM competent and confident teachers: The creation of a STEM certificate for elementary education majors. *Journal of College Science Teaching, 42*(2), 18–24.

Nadelson, L. S., Callahan, J., Pyke, P., Hay, A., Dance, M., & Pfiester, J. (2013). Teacher STEM perception and preparation: Inquiry-based STEM professional development for elementary teachers. *The Journal of Educational Research, 106*(2), 157–168.

Nadelson, L. S., Seifert, A., Moll, A. J., & Coats, B. (2012). iSTEM summer institute: An integrated approach to teacher professional development in STEM. *Journal of STEM Education, 13*(2), 69–84.

Papert, S. (1993). *The children's machine: Rethinking school in the age of the computer.* New York: Basic Books.

Papert, S., & Harel, I. (1991). Situating constructionism (preface). In I. Harel & S. Papert (Eds.), *Constructionism: Research reports and essays, 1985–1990* (pp. 1–11). Norwood, NJ: Ablex.

Park, J., Kim, D.-E., & Sohn, M. (2011). 3D simulation technology as an effective instructional tool for enhancing spatial visualization skills in apparel design. *International Journal of Technology and Design Education, 21*(4), 505–517.

Purpur, E., Radniecki, T., Colegrove, P. T., & Klenke, C. (2016). Refocusing mobile makerspace outreach efforts internally as professional development. *Library Hi Tech, 34*(1), 130–142.

Russell, A. L. (1995). Stages in learning new technology: Naive adult email users. *Computers & Education, 25*(4), 173–178.

Sheridan, K., Halverson, E. R., Litts, B., Brahms, L., Jacobs-Priebe, L., & Owens, T. (2014). Learning in the making: A comparative case study of three makerspaces. *Harvard Educational Review, 84*(4), 505–531.

Shulman, L. S. (1986). Those who understand: Knowledge growth in teaching. *Educational Researcher, 15*(2), 4–14.

Shulman, L. (1987). Knowledge and teaching: Foundations of the new reform. *Harvard Educational Review, 57*(1), 1–23.

Skaza, H., Crippen, K. J., & Carroll, K. R. (2013). Teachers' barriers to introducing system dynamics in K-12 STEM curriculum. *System Dynamics Review, 29*(3), 157–169.

Smith, S. (2014). Through the teacher's eyes : Unpacking the TPACK of digital fabrication integration in middle school language arts. *Journal of Research on Technology Education, 46*(2), 207–227.

Stager, G. S. (2013). Papert's prison fab lab: Implications for the maker movement and education design. In *Proceedings of the 12th international conference on interaction design and children* (pp. 487–490). New York, NY: ACM.

Stohlmann, M., Moore, T., & Roehrig, G. (2012). Considerations for teaching integrated STEM education. *Journal of Pre-College Engineering Education Research (J-PEER), 2*(1), 28–34.

Sun, P. C., Finger, G., & Liu, Z. L. (2014). Mapping the evolution of eLearning from 1977–2005 to inform understandings of eLearning historical trends. *Education Sciences, 4*(1), 155–171.

Tangdhanakanond, K., Pitiyanuwat, S., & Archwamety, T. (2006). A development of portfolio for learning assessment of students taught by full-scale constructionism approach at Darunsikkhalai School. *Research in the Schools, 13*(2), 24–36.

Tyler-Wood, T., Knezek, G., & Christensen, R. (2010). Instruments for assessing interest in STEM content and careers. *Journal of Technology and Teacher Education, 18*(2), 345–368.

Voogt, J., Knezek, G., Cox, M., Knezek, D., & ten Brummelhuis, A. (2013). Under which conditions does ICT have a positive effect on teaching and learning? A call to action. *Journal of Computer Assisted Learning, 29*(1), 4–14.

Chapter 6
The Emergence of the Creativity in STEM: Fostering an Alternative Approach for Science, Technology, Engineering, and Mathematics Instruction Through the Use of the Arts

Michael Marmon

Introduction

Science, technology, engineering, and mathematics (STEM) has become a ubiquitous term by which to describe the necessary skills that are essential for workers in a twenty-first century global economy. Moreover, the term has been used to equate these skills with success in both the private and public sectors. To put it more succinctly, there is the belief that students educated in STEM subjects tend to exhibit the following characteristics: "critical thinking, creativity, innovation, communication, collaboration and entrepreneurship" (Jolly, 2014 para. 1). Yet, there is an emerging movement to end the separation of science and the arts and to include the latter into the concept of STEM to further some of the aforementioned characteristics typically associated with science, technology, engineering, and mathematics.

Specifically, STEAM represents an evolution from the concept of STEM in that the inclusion of the arts is centered around stoking or bolstering the "imagination through innovation" of students as they approach STEM-related issues (Feldman, 2015 para. 4). A STEAM-centric curriculum offers an opportunity to inject creativity into courses that have traditionally been more scientific in nature. The inclusion of creativity, more specifically the arts, requires that the participating learner approach STEM activities in a distinctly different manner. Namely, STEAM establishes an intersection between disciplines while emphasizing elements of "design, performing arts (technical communication) and creative planning" (Jolly, 2014 para. 12).

This chapter seeks to fully examine the role that the arts and creativity play within the discipline of science, technology, engineering, arts, and mathematics (STEAM). This focus will investigate this role and its effectiveness within

---

M. Marmon (✉)
University of North Texas, Denton, TX, USA

© Springer Nature Switzerland AG 2019
M. S. Khine, S. Areepattamannil (eds.), *STEAM Education*,
https://doi.org/10.1007/978-3-030-04003-1_6

completing traditional STEM-related activities. To better understand the effectiveness of teaching courses with STEAM elements, student satisfaction for this approach will be both examined and discussed. This chapter will conclude with a discussion about STEAM-related theories and practices as well as the future of this emerging discipline from a technological perspective.

The Evolution of Science, Technology, Engineering, and Mathematics (STEM) to Include the Arts

The conceptual background of this chapter is the subject of STEAM with its emphasis on leveraging creativity in higher education; however, it would be beneficial to establish a foundation for this subject. The evolution of STEAM begins with its precursor, science, technology, engineering, and mathematics, colloquially known within the realm of academia as STEM. The clear difference between STEM and STEAM is that the latter extends past the apparent absence of art from the former. The prevailing definition of STEM as a concept is relatively high level in that it refers to instructional activities involving the aforementioned STEM fields occurring across all grade levels, whether the student is currently in kindergarten or getting their doctorate (Gonzalez and Kuenzi, 2012a, b).

The decision to include a radically different concept such as art within the realm of STEM acknowledges that the original field definition was lacking an important facet. The perceived value of STEM education lays within the reality of an ever-changing twenty-first economy. The belief is that there is a shortage of human capital, that is to say, individuals educated skilled in the STEM fields, where these skills are transferable to a myriad of different occupations (Marginson, Tytler, Freeman, & Roberts, 2013). The argument could be made that the inclusion of the arts and, more abstractly, creativity into STEM is to serve the noble purpose of saving the arts in education thereby putting an intrinsic value on this concept in a twenty-first century economy.

The addition of art into the realm of STEM offers a means by which to connect these concepts. White (2010) asserts that these connections enabled by the arts in STEM are essential as it facilitates the various elements within this concept to ensure that this economic future is bright.

The connections that comprise White's (2010) position are the following:

- Arts education is a key to creativity.
- Creativity is an essential component of and spurs innovation.
- Innovation is agreed to be necessary to create new industries in the future.
- New industries, with their jobs, are the basis of our future economic well-being. (para. 2).

The implication to be made from this discussion of connections afforded by the arts is a concept of value placed on the role of creativity in an economy such as this would be viewed as valuable. By placing an emphasis on innovation through art and creativity within our economy, it would lead individuals to value their presence within education given that it stimulates economic growth (White, 2010). Moreover, creativity, which was discussed earlier, is an essential element to establish innovation and, thus, is quickly being recognized as a characteristic that is desirable for a worker in this type of economy.

The role of the arts in the creative process is one of the factors that individuals cite as the foundation for perpetuating innovation in the workforce of the future regardless of the particular industry. One of the long-standing opinions for 50% of the employers surveyed in the report *Ready to Innovate*, a report compiled by Lichtenberg, Woock, and Wright (2008), is that the "arts is the most significant indicator of creativity" in the individuals that pursue positions with their firms. Furthermore, these types of individuals entering the workforce will have a hybrid mindset that will be more valuable because of their approach to problem-solving and ability to innovate new ideas because of their distinct design process. Namely, the inclusion of arts in STEM will result in individuals who view the world through a different lens that establishes a creative method by which to design processes supplemented with a whole host of technical skills as well (Feldman, 2015).

The intent of this chapter is to examine the role of STEAM in the realm of higher education instruction; however, to fully examine this subject, it is required to understand the evolution of STEM into STEAM. Moreover, by examining the role that STEM has on the economic prospects of learners and the purpose of higher education to construct/inform an individual's viability in an economy, a discussion of the role of creativity and the arts is essential. The clarity of how the arts fit within STEAM provides a necessary implication about its purpose and value within the economy of both the present and future for graduates who possess these skills while also asserting its need within higher education.

The Emergence of Science, Technology, Engineering, Art and Mathematics (STEAM) in the Field of Higher Education

The evolution of STEM to STEAM provided a brief explanation in the relevance of the arts within the traditional STEM fields of science, technology, engineering, and mathematics. This discussion yielded some context about the value of art and by extension creativity in an economy that many observers feel will be shaped by the intellectual sphere of STEM disciplines. This chapter will provide a more thorough examination into the concept of STEAM and creativity within the context of higher education to better understand the ability of art to transform the thought processes typically associated with a STEM curriculum. To accomplish this examination and the resulting understanding of creativity within the arts, a discussion will occur in this section that highlights STEAM learning programs and instructional design theories.

A short definition for the concept of STEAM was provided in the previous section that spoke of the term from a perspective of its purpose; thus, it is beneficial to provide a more technical definition that provides the theoretical underpinning of this concept. The term of STEAM originated in 2010 and is attributed to Harvey White, a founder of Qualcomm, who felt that the STEM field could benefit from the "habits of mind of the artist" with an emphasis "creative problem solving" cultivated from a defined approach to instruction (STEAM Programs, n.d.). The position of Harvey White as the preeminent proponent of STEAM provides the foundation to what the concept would evolve into years later. STEAM within the context of this chapter will be defined through the intersection of STEM and the function that the arts will play within this emerging discipline.

Interjecting the Arts into STEM to Create an Interdisciplinary Education

STEM as a concept refers to the "integration of science, technology, engineering and mathematics in a new cross-disciplinary subject in schools" (Dugger, 2010 p.2). The working definition of STEM provided by Dugger leads one to wonder how the arts can integrate within the structure of these four fields. The answer to this thought is that the arts act as a connector to all of these concepts, which results in an understanding about how each can be leveraged and their purpose in society. In particular, the arts acts as the method by which society is able to construct, convey, and comprehend "attitudes and customs in the past, present and future" (Yakman, 2008 p.16). The position of Yakman offers a context for how the arts can be utilized to present complex STEM-related concepts in a way that is understood by individuals who are not pursuing a degree within a major from one of those fields. It is with this thought in mind that provides the perspective by which to understand how the arts and creativity are used within traditional STEM-centric higher education programs.

The foremost aspect of STEAM in higher education is the fulfillment of an important concept within the realm of higher education, a well-rounded and comprehensive approach to learning that is interdisciplinary in nature. The relevance of an interdisciplinary STEM curriculum that has been infused with the arts is that students are taught concepts and information from the arts and engineering courses that are relevant to the students pursuit of a degree in either discipline. An example of this notion is engineering students completing an "arts or drafting course" to understand the "practical aspects of perspective in drawing and the structural elements of construction" (Robinson & Baxter, 2013 p. 3). The interesting element of Robinson and Baxter's example is that it highlights a recurring belief discussed throughout this chapter; thus far, the presence of arts courses within a traditional STEM curriculum will establish a new method by which to visualize and approach their work in the said courses. In particular, the position of Robinson and Baxter highlights the belief of commonality or common processes that could be utilized by students in either of these programs. For instance, individuals within a science program seek to understand the inner workings of the world through the collection of data, whereas arts approach their creative works by replicating the world through similar visual observances (Fulton & Simpson-Steele, 2016).

The principles that Fulton and Simpson-Steele are proposing is that commonalities in processes emphasized in a STEAM curriculum or a course that is based on either science or the arts minimizes the barriers to constructing knowledge from a practical perspective. The connection of these common elements in the mind of the learner will emphasize "engagement with ideas rather than teaching of concepts" (Fulton & Simpson-Steele, 2016 p. 10–11). The purpose of the emphasis on the ideas rather than the method of instruction implies that a student would be receptive to how the content can be utilized to support their own research interests as opposed to approaching it from an specific academic perspective. For instance, there is less value for individuals to have extensive memories comprised of mathematic equations or scientific information as technology such as computers can provide this information; thus, creativity offers a method by which to solve questions of this type (Callahan, 2014).

Leveraging Connections and Information in STEAM Courses to Improve Learning Outcomes

It is important to note that the role of creativity and the arts as means to establish the connections by which to resolve complex issues in a STEAM-centric curricula is through the construction and utilization of information. This approach to constructing knowledge in STEAM courses offers a glimpse by which to understand the improvement of a STEM curriculum through the learning outcomes for the individuals completing these classes. A study conducted by Mishra and Henriksen (2013) found that these STEAM instructional methods yield "more motivated and engaged students" that were capable of increasing their "learning within these disciplines" (p. 4). While all instructors seek to ensure that their learners are achieving positive learning outcomes, it is necessary to delve further into the subject of these outcomes and how these artistic/creative instructional methods have improved upon traditional STEM courses.

While the discussion of critical thinking has been discussed previously in this chapter, it would be beneficial to delve into the concept further as it relates to STEAM. The ability for a learner to process their thoughts creatively results in an enhanced ability to solve the problems presented to them which is a required skill in higher education courses (Art, 2013). The prevailing belief to be gleaned from this assertion speaks to the inclusion of creativity resulting in learners becoming "better problem solvers," who are able to better understand these subjects as a result (Art, 2013 para. 8). The question that arises from statements such as this is the validity of the claim that lie within them. The answer to this question resides within the research findings of studies conducted about the influence and effectiveness that the arts and, by extension, creativity have within STEAM education.

There have been a multitude of studies that have sought to better understand the impact that artistic elements have within higher education courses. In particular, these studies have found that the influence of creativity on learners results in higher level thought, which include an ability to navigate difficult and stressful situations, in addition to enhanced life and social skills (Autry & Walker, 2011; Clinton & Hokanson, 2012; Hargrove, 2012). As one can ascertain, there is a clear and positive impact on learners that extend past the classroom. While this is certainly a positive aspect from integrating creativity in a course, it is best to refocus onto its role in a STEAM context.

The context of STEAM in a twenty-first century world is predicated on the ability to leverage creative solutions to the problems that are facing individuals living at this particular moment in history. Creativity, in its most simple or mundane form, has the opportunity to change how the layperson sees critical problems such as climate change or controversies involving vaccinations through the framing of data obtained through research (Thurley, 2016). The assumption to be made is that the "layperson" will obtain these skills from a higher education experience that is centered around STEAM. Thurley (2016) proposed the value of instilling creative endeavors such literature or creative writing in academic communication or scholarly articles which will result in findings that are easier to understand by those unfamiliar with the subject. Granted, the proposal offered by Thurley is one of a multitude that could be discussed in many volumes; thus, the focus of this chapter will turn from potential applications of STEAM in higher education to student satisfaction in these courses.

Determining the Impact of STEAM by Examining Student Satisfaction in These Courses

The emergence of STEAM as a method by which creativity and the arts are utilized in a course has an impact on the level of student satisfaction. Namely, these courses could be extraordinarily designed by an instructor or instructional designer, but if the content and processes do not resonate with the participating students, then there is a problem with the course's impact or effectiveness. With this in mind, it is beneficial to better understand how STEAM-centric courses are received by the students that have participated in them. Moreover, there is value in discussing the methods that STEAM instructors find to be particularly effective and tends to evoke the most from their students from the perspective of learning outcomes.

STEAM and Student Satisfaction

The primary method by which it is possible to measure student satisfaction in a STEAM curriculum is through the course elements and structure that comprise these courses. One such method is to present the content in such a way that it is possible to motivate learners through an "inquiry-based approach to learning," where knowledge is constructed through an experience shaped by creativity and coalesced from a "broader understanding of all the parts" of the subject (Helfferich, Dawe, & Tarnai, 2014 p. 2). To obtain the insights necessary about student satisfaction with regard to the inquiry nature of these STEAM courses, satisfaction surveys assume the foremost method by which to evaluate the sentiment levels of each learner within these courses. The results obtained from these surveys will not only provide insights into the feelings of individuals within these courses; they will offer a means by which to measure the level of success for the STEAM course as a whole (Advancing, n.d.).

Increasing Student Satisfaction Through a Framework That Engages and Challenges Students

As satisfaction surveys offer a means by which to evaluate the success of the learners participating in a higher education STEAM course, it is necessary to understand how to shape the emotional and intellectual response in the mind of the participants. Arnold and Reeves (2014) highlight the necessity of an instructor to develop a framework that is "effective" in "increasing student retention and persistence" in these courses through the constant engagement of the learner (p. 2). The type of framework that should be designed for a STEAM course should leverage the environment that the course takes place in (online or face-to-face) and utilize diverse activities that lend themselves to advancing the knowledge through the aforementioned active learner engagement (Lo, 2010). After all, it should be logical that the challenge and academic rigor of the activities within these courses would accurately reflect the satisfaction of the learner as these activities stimulate their desire to learn and participate.

The approach or framework for designing an experience in STEAM courses should be multidisciplinary in nature, which means that it fits the core ideals of STEAM and include within its structure activities found in the real world. One of the prominent frameworks in higher education is the teacher education program at the State University of New York at Potsdam that has taken this multidisciplinary message to heart and developed a curriculum that engages learners with examining concepts from the arts and sciences and developing real-world problem-solving skills from the information in these courses (Madden et al., 2013). The Potsdam model for the use of STEAM in higher education is fascinating from both its philosophical foundation and its practical implementation of these ideas that resulted in a comprehensive degree program that leverages content from a multitude of diverse academic programs.

The implementation of this framework is predicated on the idea of developing a curriculum centered around the notion of STEAM theory and its practice, which requires a measure of cooperation between stakeholders in the academic departments creating these courses. This cooperation between these individuals requires extensive communication as it is illogical for a subject matter expert in one academic discipline to have an expertise in another STEAM field as well. After all, a STEAM course represents a multidisciplinary approach for instruction, and each of these elements should not be presented to students as separate concepts or content elements but rather be treated as one concept highlighting the connections between each element (Land, 2013). This collaboration between course stakeholders as explained by Land is a necessary activity that also yields an essential value for the program by two different departments and makes them more likely to participate in conjunction with each other outside of a single STEAM course.

The SUNY Potsdam model is one specific example of a dual degree program that is comprised of the arts and other STEAM disciplines. The purpose of this type of dual degree program is the comprehensive nature of the curriculum that is not found in a single STEAM course that exists in a vacuum by itself. Thus, these courses are not merely offering an isolated requirement on a degree plan; it becomes an example for the previously discussed notion about the economic value of these STEAM courses. More appropriately, these dual degree programs allow difficult content types to transition from STEM-related skills into "implemented and fielded capabilities which require more creative skills" (Land, 2013 p.552).

Most importantly, a curriculum centered around STEAM courses and concepts would also have the added benefit of a more comprehensive design rather just being a single course that fits as an elective in another program. Specifically, programs such as this one take an approach that not only sets its own cumulative learning outcomes, it will approach the construction of knowledge within a program of this type progressively and will be addressed by benchmarks or artifacts such as portfolios for learners (Madden et al., 2013). Ultimately, the benefits of a STEAM framework and curriculum such as this in higher education plays to the advantage of the student as it provides a comprehensive experience that not only build upon itself, it also presents a multitude of similar concepts within one academic department rather than across a myriad of other ones, thereby making it easier for them to access these courses.

Charting the Future of Creativity and Arts in a Forthcoming STEAM Course

The academic discipline of STEAM is not one that remains static from either a conceptual or theoretical perspective, it will continue to evolve and remain applicable well into the future as it is centered around the notion of innovation. Moreover, creativity exists as a means of critical thinking to facilitate the type of innovation required to resolve the problems of the future. The future of STEAM education has to be as flexible and timely as the skills and expectations associated with them. These curriculum decisions must be made by isolating the technologies/skills required to fulfill a particular educational and economical need of a society at that particular moment in human history. This is the foundation for the future of STEAM education, the ability to evolve and create the necessary connections from a conceptual and theoretical context.

The Foundation of STEAM's Future Resides with Learner Engagement and Instructional Design Processes

The future of STEAM education resides within its ability to attract and engage students in courses that leverage instructional design elements in such a way that the learners can relate the content to a real-world context. This relationship of STEAM concepts, technologies, and instructional theories will result in a comprehensive knowledge that can best be described as a functional literacy. More appropriately, learners completing these STEAM courses will be capable and confident enough to draw the connections between each of these elements, thereby enabling them to evolve intellectually within their chosen profession through the act of observation, critical thought, and action as required (Yakman, 2008). The novel aspect of a functional literacy in STEAM is not merely the knowledge acquired on an individual subject but rather the ability to creatively utilize it in an ever-changing economic or social environment.

In particular, this functional literacy is a direct result of approaching curriculum development with the holistic mindset discussed earlier, which lends itself to learner's engagement through the method of instruction and the furthering of the learner's intellectual capacity. The implication here is that STEAM education seeks to develop individuals dedicated to the value of lifelong learning through engaging their sense of "logical thinking and problem-solving abilities" borne from the aforementioned holistic learning rather than the "fragmentation of knowledge and memorization processes" (Developing, n.d., p. 3). The benefit of a holistic STEAM curriculum is the knowledge acquisition that is approached from a collective position rather than merely a recitation of facts, which will result in an understanding of the content from a natural perspective. The next aspect of the future of STEAM education is technology, which aids in engaging the learner and providing a method by which to apply the knowledge acquired in a course of this type.

Shaping Critical Thought Processes with New Media Literacies

It is possible to build upon the importance of learner engagement by either developing or utilizing emerging technologies within a STEAM course to cultivate the learner's interest or even facilitate their ability to participate in these courses as well. The foremost method by which to engage the learner's interest is to incorporate new media literacies in these courses as they connect to the economy that they will enter upon graduating from an institution of higher learning. The cost of implementing these new media literacies is relatively low for the instructor as most students have the ability to transfer their thoughts and research into a visual mode (Land, 2013). Land (2013) further explains that these foundational skills to translate concepts from one medium to another will occur by incorporating "performance, simulations and collective intelligence" into their courses via prompts (p. 550).

The area of expansion afforded by new media literacies is the relationship between an environment and technology but the experiences that result from the said relationship as well. There is a practical example for leveraging new media literacies and technologies to present information in ways that are more accessible for individual learners (Tsoupikova, Silva, Kostis, & Shah, 2014). It is this convergence of these essential elements (new media literacies and technologies) that results in a learning experience that produces authentic inquiry-based learning. Tsoupikova et al. (2014) explain that museums offer an example environment for this transfer of knowledge between mediums to establish the connections required for inquiry-based learning through the utilization of "augmented reality, challenges and adventure games" (para. 5–6). New media literacies are going to be an essential element for creating learning experiences as they conjure new methods by which to present content and information to learners in an engaging manner. As was mentioned earlier, the interplay between new media literacies and the environments that house them will be guided in the future by the technologies that are currently emerging within the field of STEAM.

The Utilization of Emerging STEAM Technologies

The position of Land (2013) offers a logical segue from a new media literacy perspective as its relate to STEAM, as physical technologies will not only engage the intellect of the participating learner, it also affords the students with another avenue by which to develop their respective knowledge of the content and a physical artifact as well. The emerging digital technologies that continue to evolve rapidly for learners include "artificial intelligence, DNA mapping, robotics, nanotechnology, 3D printing, biotechnology and the 'Internet of things'" and offer a variety of creative avenues by which to creatively develop responses to real-world problems (Taylor, 2016 p. 90). This exposure to multiple technologies with creative applications necessitates an understanding of how to utilize them in such a way that is not only creative but also functions outside the realm of a classroom setting. More importantly, this real-world application of the technology requires that the learners understand that these technologies are utilized by their peers in similar contexts (Lewontin, 2015). In this sense, the practical application of technology increases the learner's engagement with these tools through utilization, and thus, there is intrinsic value in leveraging them in a variety of different environments.

After giving insight into role of technologies and creativity from a theoretical perspective, it is possible for one to discuss the emerging tools in the discipline of STEAM. While there is a multitude of current and emerging technologies to examine, it is more beneficial to draw connections between a selected few technologies and the creative process as opposed to merely highlighting a list of various devices. One of the previously mentioned technologies that will become more prevalent in STEAM education because of its practicality and influence on the creative design process is 3D printing. The usage of a 3D printer within a STEAM course is twofold; the first aspect is its influence on the creative process from the perspective of design and the second is that these devices provide a tangible artifact that reflects the knowledge obtained from the activity (Lonka & Cho, 2015). The importance of this practical application of creativity and a design process results in a greater level of learning that occurs as the learner is capable of connecting the technology to the presented concepts in a realistic situation (Lonka & Cho, 2015). This is an interesting perspective as it implies that the technology acts as a method by which to facilitate or influence the application of knowledge gleaned in a course by guiding the creative thoughts of the learner.

Another emerging technology that contains builds upon this notion of facilitating the relationship between creativity and knowledge is augmented reality (AR). The basic premise of augmented reality is that it is meant to combine virtual elements in a real-world context. A more accurate definition is the "coexistence of virtual objects and real environments," which affords the learner or more abstractly the user the opportunity to learn within context in such a way that enhances their sense of reality (Wu, Lee, Chang, & Liang, 2013). Even within the realm of augmented reality's definition, it is possible to understand its application within STEAM education and the role of creativity in the design of these AR systems. The foremost benefit of this technology is that it increases the engagement of the learners that develop and utilize it within these courses through the process of active inquiry learning (Ahn & Choi, 2015). This crux of this technology and its relevance to creativity and the arts lay with its focus on creative design from a myriad of positions. Specifically, design begins with the "information and system quality," which refers to how the system will function to the aesthetics and a learner experience that deals with "the visual design and the physiological reactions" of the participant as well (Huang & Liao, 2015 p. 275). Huang and Liao (2015) highlight the undercurrent of STEAM within the context of augmented reality through the distinct relationship of system design (from a technology perspective) and creative design (aesthetics and user experience) working in concert to develop applications that would be for the benefit of society.

Technology represents an essential element within the realm of STEAM education, and a survey of the gamut of emerging technologies goes beyond the scope of this chapter; however, the discussion of a few promising technologies assists with understanding of the discipline's future. This section focused on two particular STEAM technologies, 3D printing and augmented reality, both of which allow for a connection between creativity and the other STEAM fields. The defining aspect of these technologies is not the technology itself nor the benefits of its usage; it is the inherent nature of creativity afforded by them. It is evident that both 3D printing and augmented reality lend themselves quite extensively to the notion of the creative design process as discussed throughout this chapter. This exemplifies the purpose for including two such emerging technologies in this chapter as they offer a reference for a "functional design process" that ultimately informs the "aesthetic nature and utility of items" to resolve the problems that lay in front of the learner (Bequette & Bequette, 2012 p. 40).

Conclusion

The conceptual foundation of this chapter was the investigation into the role that creativity and the arts play within the emerging field of science, technology, engineering, art, and mathematics or STEAM. STEAM is borne from the principles of STEM with the added facet of integrating the arts into these courses, which offers instructors within higher education the opportunity to enhance the creative thinking and problem-solving abilities of their students. As humanity continues to progress through the twenty-first century, the economy has become centered around the traditional STEM fields and is quickly becoming a sought after and essential collection of skills for one navigating the future of our species. In particular, it has been theorized that these necessary "science and technology-based innovation" skills are required in the industries of the twenty-first century as they would be "impossible without a workforce educated in science, technology, engineering and math" (Atkinson & Mayo, 2010 p. 21).

While STEM-related skills are essential within this economy of the future, individuals began to notice the applicability of establishing the arts within the said skill sets. It has been stated earlier in this chapter that the arts offer a means by which to further innovation and creative thinking through establishing connections between disciplines. Specifically, STEAM provides the means to "connect disciplines that were previously perceived as disparate" and serves the purpose of "enhancing student interest and showing the value" in investigating these concepts from an interdisciplinary perspective (Guyotte, Sochacka, Costantino, Walther, and Kellam, 2014 p. 12). The intent of STEAM within higher education is to further the innovation "demanded by the 21st Century" and the addition of the arts to STEM endows the learner with the design skills necessary to "create the innovative products and solutions that will propel our economy forward" (Maeda, 2013 p. 1). The position of Maeda provides the necessity of halting the notion that the arts should remain separate from the fields of science and mathematics.

By establishing the rationale and need for adding the arts to STEM from an economical perspective, it is possible to entice the skeptical into understanding the economic promise of STEAM in the future. This understanding of the purpose and potential of STEAM will lead to increased enrollment in these courses, as well as graduates that are well-rounded both academically and intellectually. Throughout the course of this chapter, there was a discussion as to this impact of the arts and

6 The Emergence of the Creativity in STEM: Fostering an Alternative Approach…

creativity on STEM that reinforces the belief that these elements impact the creative thinking and problem-solving of the individuals participating in a STEAM curriculum.

It is through this understanding of the elements and activities that construct a STEAM course, which influence how the courses are received by the students that complete them. Student satisfaction in these courses is important because it is an active reflection on the instructional design theories performed in practice and the effectiveness of these choices that determine how the students respond to a STEAM course. The discussion of this topic during this chapter revealed that the implications of student satisfaction rely on a positive influence on the continued presence of the learner in these courses and their success in these courses as well.

STEAM represents the future of creativity and innovation within both a twenty-first century global economy and higher education, where individuals need the skills necessary to navigate these complex disciplines and intellectual concepts. Moreover, the presence and emphasis on the creativity afforded through the inclusion of the arts in STEM make it possible to achieve the following notion: the cultivation of learners who "approach STEM subjects creatively and make them real-world-relevant" (Feldman, 2015 para. 9). If we as a society continue to separate the arts from science and mathematics, we will not only be putting our learners at a disadvantage, we are putting the future of our species as whole at one as well. STEAM and creativity is our best chance to produce well-rounded learners and by extension workers that are capable of innovating in a multitude of disciplines into the twenty-first century and beyond.

References

Advancing Student Learning. (n.d.). Columbia University. Retrieved November 29, 2018, from http://www.columbia.edu/cu/provost/midstates/docs/AdvancingSL.pdf

Ahn, H. S., & Choi, Y. M. (2015). Analysis on the effects of the augmented reality-based STEAM program on education. *Advanced Science and Technology Letters, 92*, 125–130.

Arnold, B., & Reeves, J. (2014). Translating best practices for student engagement to online STEAM courses. In *Proceedings of the 2014 American Society for Engineering Education Zone IV Conference, Long Beach CA*.

Art could help create a better 'STEM' student. (2013, December 3). Retrieved March 1, 2017, from https://www.sciencedaily.com/releases/2013/12/131203091633.htm

Atkinson, R. D., & Mayo, M. J. (2010). Refueling the US innovation economy: Fresh approaches to science, technology, engineering and mathematics (STEM) education.

Autry, L. L., & Walker, M. E. (2011). Artistic representation: Promoting student creativity and self-reflection. *Journal of Creativity in Mental Health, 6*(1), 42–55.

Bequette, J., & Bequette, M. B. (2012). A place for art and design education in the STEM conversation. *Art Education, 65*(2), 40–47.

Callahan, D. (2014). The importance of being creative. *The STEAM Journal, 1*(2). https://doi.org/10.5642/steam.20140102.4. Available at: http://scholarship.claremont.edu/steam/vol1/iss2/4. Accessed 12 Feb 2017.

Clinton, G., & Hokanson, B. (2012). Creativity in the training and practice of instructional designers: The Design/Creativity Loops model. *Educational Technology Research and Development, 60*(1), 111–130.

Developing STEAM Education to Improve Students' Innovative Ability. (2016, October 10). Retrieved November 29, 2018, from https://steamedu.com/developing-steam-education-to-improve-students-innovative-ability/

Dugger, W. E. (2010). Evolution of STEM in the United States. In *6th Biennial international conference on technology education research, Gold Coast, Queensland, Australia.*

Feldman, A. (2015). Why we need to put the arts into STEM education. Retrieved March 1, 2017, from http://www.slate.com/articles/technology/future_tense/2015/06/steam_vs_stem_why_we_need_to_put_the_arts_into_stem_education.html

Fulton, L. A., & Simpson-Steele, J. (2016). Reconciling the divide: Common processes in science and arts education. *The STEAM Journal, 2*(2), 3.

Gonzalez, H. B., & Kuenzi, J. J. (2012a). *Science, technology, engineering, and mathematics (STEM) education: A primer.* Congressional Research Service, Library of Congress.

Gonzalez, H. B., & Kuenzi, J. J. (2012b). *Science, technology, engineering, and mathematics (STEM) education: A primer.* Congressional Research Service, Library of Congress.

Guyotte, K. W., Sochacka, N. W., Costantino, T. E., Walther, J., & Kellam, N. N. (2014). STEAM as social practice: Cultivating creativity in transdisciplinary spaces. *Art Education, 67*(6), 12–19.

Hargrove, R. (2012). Fostering creativity in the design studio: A framework towards effective pedagogical practices. *Art, Design & Communication in Higher Education, 10*(1), 7–31.

Helfferich, D., Dawe, J., Tarnai, N. (2014). STEAMpower: Inspiring students, teachers, and the public. AFES Miscellaneous Publication MP 2014-13. Retrieved February 16, 2017.

Huang, T. L., & Liao, S. (2015). A model of acceptance of augmented-reality interactive technology: The moderating role of cognitive innovativeness. *Electronic Commerce Research, 15*(2), 269–295.

Jolly, A. (2014). STEM vs. STEAM: Do the arts belong. *Education Week.* p. 18.

Land, M. H. (2013). Full STEAM ahead: The benefits of integrating the arts into STEM. *Procedia Computer Science, 20*, 547–552.

Lewontin, M. (2015). How efforts to combine arts with STEM education could improve tech diversity. Retrieved March 1, 2017, from http://www.csmonitor.com/Technology/2015/1216/How-efforts-to-combine-arts-with-STEM-education-could-improve-tech-diversity

Lichtenberg, J., Woock, C., & Wright, M. (2008). *Ready to innovate: Are educators and executives aligned on the creative readiness of the US workforce?* New York: Conference Board.

Lo, C. C. (2010). How student satisfaction factors affect perceived learning. *Journal of the Scholarship of Teaching and Learning, 10*(1), 47–54.

Lonka, K., & Cho, V. (2015). *Innovative schools: Teaching & learning in the digital era (European parliament, directorate-general for internal policies, culture and education).* Brussels, Belgium: Policy Department – European Union.

Madden, M. E., Baxter, M., Beauchamp, H., Bouchard, K., Habermas, D., Huff, M., et al. (2013). Rethinking STEM education: An interdisciplinary STEAM curriculum. *Procedia Computer Science, 20*, 541–546.

Maeda, J. (2013). STEM+ Art= Steam. *The STEAM Journal, 1*(1), 34.

Marginson, S., Tytler, R., Freeman, B., Roberts, K. (2013). STEM: Country comparisons: International comparisons of science, technology, engineering and mathematics (STEM) education. Final report.

Mishra, P., & Henriksen, D. (2013). A new approach to defining and measuring creativity: Rethinking technology & creativity in the 21st century. *TechTrends, 57*(5), 10.

Robinson, C., & Baxter, S. (2013). Turning STEM into STEAM. *Age, 23*, 1.

STEAM Programs. (n.d.). Retrieved March 1, 2017, from http://uwl-web.com/integrated/?page_id=108

Taylor, P. C. (2016). Session N: Why is a STEAM curriculum perspective crucial to the 21st century?

Thurley, C. W. (2016). Infusing the arts into science and the sciences into the arts: An argument for interdisciplinary STEAM in higher education pathways. *The STEAM Journal, 2*(2), 18.

Tsoupikova, D., Silva, B., Kostis, H., Shah, T. (2014). Girls steaming to practicing STEAM: The powers of new media arts for engaging girls in STEM. Retrieved March 1, 2017, from http://median.newmediacaucus.org/caa-edition/girls-steaming-to-practicing-steam-the-powers-of-new-media-arts-for-engaging-girls-in-stem/

White, H. (2010). Science, technology, engineering, art, mathematics. Retrieved March 1, 2017, from http://steam-notstem.com/

Wu, H. K., Lee, S. W. Y., Chang, H. Y., & Liang, J. C. (2013). Current status, opportunities and challenges of augmented reality in education. *Computers & Education, 62*, 41–49.

Yakman, G. (2008). STEAM education: An overview of creating a model of integrative education. In *Pupils' attitudes towards technology (PATT-19) conference: Research on technology, innovation, design & engineering teaching, Salt Lake City, Utah, USA.*

Chapter 7
Developing a Rhetoric of Aesthetics: The (Often) Forgotten Link Between Art and STEM

Rohit Mehta, Sarah Keenan, Danah Henriksen, and Punya Mishra

Euclid alone has looked on Beauty bare.
Let all who prate of Beauty hold their peace,
And lay them prone upon the earth and cease
To ponder on themselves, the while they stare
At nothing, intricately drawn nowhere

— Ed St. Vincent Milay

The greatest scientists are artists as well

— Albert Einstein

Introduction

A child's first experience, of peeking through a telescope to see the vivid sharply etched, yet fragile, rings of Saturn is a powerful one; perhaps as powerful as standing amidst redwood trees listening to the sound of wind rustling through the leaves or experiencing a moment of clarity when an elegant geometrical proof, surprising in its simplicity, emerges from a chaos of sketches and doodles. It is in this sense of awe and wonder that our minds nibble at confronting powerful *ideas* such as infinity (whether the infinity of numbers, or the interminably large scale of the cosmos, or the immeasurably small universe of cells and atoms and quarks). The emotional turbulence that overwhelms us when we reflect on nature, truly understand a scientific idea, or solve a tricky mathematical or engineering problem often leads to powerful aesthetic experiences. These experiences, we argue, are no different or less

---

R. Mehta (✉)
California State University, Fresno, CA, USA
e-mail: mehta@csufresno.edu

S. Keenan
Michigan State University, East Lansing, MI, USA

D. Henriksen · P. Mishra
Arizona State University, Tempe, AZ, USA

© Springer Nature Switzerland AG 2019
M. S. Khine, S. Areepattamannil (eds.), *STEAM Education*,
https://doi.org/10.1007/978-3-030-04003-1_7

than the aesthetic experience we have in engaging with powerful artistic human creations, be it music or the visual arts.

Our case for marrying the arts with STEM into a STEAM view of learning pivots on such aesthetic experiences of beauty, curiosity, wonder, awe, and the inherent pleasure of figuring things out. In this chapter, we put forward our rationale for how aesthetic experiences are the often forgotten link between the arts and STEM. Given this link, we propose developing a *rhetoric of aesthetics* in STEM as a practice-based approach to implementing STEAM-based teaching and learning. We share outcomes of our own research which led to the development of a threefold rhetoric that explains the role of the aesthetic in STEM disciplines. The three frames that emerged from this work involve intersections of arts and STEM and can, therefore, be seen as the fuel to designing STEAM pedagogies. Finally, we give examples of how we have used this rhetoric to guide teacher professional development for STEM educators, focused on building a more aesthetically driven STEAM view of learning.

Rationale for the Rhetoric

A rhetoric of aesthetics in STEM emerges from examples of how scientists, mathematicians, and engineers explain and understand their lived experiences of doing science and mathematics. They often speak in affective terms, of beauty and elegance—of frustrations at momentary failures and pleasure at the process or the culminating, momentous thrill of discovery (e.g., Hoffmann, 1990; Holton, 1988; Tauber, 1997). Through this they provide us a glimpse of an aesthetic lens that influenced their vision.

Both Hegel and Kant considered aesthetics to be a bridge between the sensuous and the intellectual (Scruton, 1983). Metallurgist, Cyril Stanley, exemplifies this connection, explaining, "The stage of discovery [is] entirely sensual and mathematics [is] only necessary to be able to communicate with other people" (as quoted by Root-Bernstein, Bernstein, & Garnier, 1995, p. 133). The deeply personal experience within scientific practice, which is often kept separate from the public face of science, is a theme repeated across the stories told by many scientists and mathematicians (as seen in autobiographical accounts, see Chandrasekhar, 1987; or for other scholars' reflections on their processes, see Girod, 2007, Tauber, 1997). However, when presenting science to the public, this "personal" account of science is often suppressed in favor of more straightforward linear accounts (Holton, 1988).

The literature around science is full of examples of beauty and wonder-driven approaches (Chandrasekhar, 1987; Orrell, 2012). These are abundant in the stories of everyday scientists and mathematicians in their discovery of the "truth" that nature holds (Girod, 2007; Mehta, Mishra, & Henriksen, 2016; Mishra, Terry, & Henriksen, 2013). From Pythagoras to Kepler, from Newton to Einstein, most stories of real-world scientific inquiries by notable scientists are laden with experiences of wonderment with inquiry (Orrell, 2012). This pairs with an awareness of beauty or elegance, both in nature, and the theories and formulae that describe nature. For instance, in an article in *Scientific American*, Nobel laureate and physicist, Paul Dirac (1963), suggested that beauty may even be the most important thing of all:

> [I]t is more important to have beauty in one's equations than to have them fit experiment … It seems that if one is working from the point of view of getting beauty in one's equations, and if one has really a sound insight, one is on a sure line of progress. If there is not complete agreement between the results of one's work and experiment, one should not allow oneself to be too discouraged, because the discrepancy may well be due to minor features that are not properly taken into account and that will get cleared up with further developments of the theory. (p. 47)

Mathematicians of course, maybe more so than scientists, speak to beauty, elegance, and simplicity in what they value in their work. Ignoring this aspect of their work, in the words of Poincare (1910), "would be to forget the feeling of mathematical beauty, of the harmony of numbers and forms, of geometric elegance. This is a true esthetic feeling that all real mathematicians know, and surely it belongs to emotional sensibility" (p. 331).

Aesthetics is by no means constrained to the realm of science and mathematics. Engineers and computer scientists speak of their work in aesthetic terms as well. This may appear somewhat surprising given that the ultimate goal of engineering is functionality—i.e., creating a functioning, working device or program or artifact. But, as Gustave Eiffel, of the eponymous tower, said, "Can one think that because we are engineers, beauty does not preoccupy us or that we do not try to build beautiful, as well as solid and lasting structures? Aren't the genuine functions of strength always in keeping with unwritten conditions of harmony?" (as quoted by Petroski, 2011). Similarly, in looking at the aesthetic aspects of computer programmers, Good, Keenan, and Mishra (2016) wrote that:

> Professional programmers will readily share their experiences with code that they might describe as elegant, beautiful, or clean… both novices and experts describe [code] as *ugly* and *beautiful*. Both groups reported aesthetic experiences related to code they had observed, albeit with functionality being of higher importance. (p. 315)

We could provide hundreds of such examples to demonstrate the significant role the aesthetic plays in how scientists engage in the STEM disciplines. But the overall point we seek to convey in this chapter is clear: *if we want students to have authentic experiences with learning in the STEM disciplines, to cultivate a true STEAM view of learning, we cannot ignore the role of the aesthetic.* Doing so does a disservice to the very reasons that *practitioners* view as key to their own motivations and passions for doing science, mathematics, and engineering. Ignoring the aesthetic, moreover, devalues the humanistic and artful aspects of the STEM disciplines, the tissues and sinew of the body of science, leaving behind just the functional and the practical, the bare bones, as it were.

Instrumental and Aesthetic: The Yin and Yang of Motivations for Science

There are a multitude of reasons that much conventional teaching and learning has often ignored the value of the aesthetic in how science and technology are represented in our curricula and presented to students in our classrooms. We suggest that in attempting to find efficient ways of teaching STEM in schools, over the years, our teaching and curricula have increasingly adopted purely functional, *instrumental* reasons for studying science, mathematics, engineering, and technology. Such instrumental reasoning positions science as a tool rather than a personally motivating, enjoyable, or beautiful subject. This happens when we push participation in STEM fields, because they may offer a high-paying profession or stable job, or because it feeds into a patriotic rhetoric wherein STEM helps nations best others in international competitions. This logic was reified in the US-USSR space race in the 1960s, where STEM was used as a tool to win at an international competition—space exploration being its by-product. This sense of war and competition as the true value of education can be seen in influential reports, such as 1983's *A Nation at Risk*, where the authors begin by stating, "If an unfriendly foreign power had attempted to impose on America the mediocre educational performance that exists today, we might well have viewed it as an act of war" (National Commission on Excellence in Education, 1983, p. 5).

Though we do not deny the value of instrumental approaches (no doubt STEM fields play a significant role in the economic growth and innovation of society), we do believe that a singular focus on such approaches misses what is engaging and motivating about science and engineering—and ironically misses the key driver that has motivated the most innovative STEM thinkers through history. We argue that, at its core, the sensation of wonderment, the sublime feeling of awe, the natural sense of curiosity, and the intrinsic joy of discovery—the *affective and emotional components* of the experience of doing science—are the key to learning in the STEM disciplines. As educators, we often ignore them at the risk of alienating the very students we want to reach. We argue that doing science is an inherently emotional, and thereby humanistic, aspect of our lives. It is fueled by curiosity, steered by wonder, soothed by beauty, and replenished by the joy of discovery. This is why we love to solve problems, explore new lands and seas, and build enormous bridges and miniscule nanobots. It is who we are as humans—curious, complex, and forward-looking. This is the aesthetic and affect-based reason for doing science. Certainly, an instrumental reason complements it by grounding us in pragmatism, but it is not the primary motivator and therefore cannot be the single, sole lens through which we view STEM pedagogy.

To be more specific, the instrumental arguments for the most part are somewhat removed from the humanizing, rich, personal, subjective, here-and-now experiences of learners. Instrumental arguments are grounded in meeting an abstract, and essentially unknowable, possible future need, rather than a tangible, concrete present. It suggests that learning the STEM disciplines would ultimately, in the future tense, be extrinsically beneficial to both the individual (e.g., successful career options, financial stability) and society (e.g., technological progress, economic growth, global competitiveness). Learners who start hearing these arguments at a young age are expected to accept these at face value. But it is not surprising that these distant goals cannot sustain motivation in STEM, if they motivate at all. We need more proximate goals that are inherently motivating and which can be infused within a STEAM curriculum.

Why the Aesthetic Is Sexy: One Possible Answer

Our previous discussion suggests that if we are to go with what motivates practicing scientists, mathematicians, and engineers, we need to consider the role of aesthetics in our curriculum. But this also begs the question of why fields focused on understanding the world or constructing things are so motivating. Why *do* we get a frisson of pleasure when we solve a problem or understand something new?

One possible answer emerges from a classic paper written by psychologist Alison Gopnik titled, "Explanation as orgasm and the drive for causal knowledge: The function, evolution and phenomenology of the theory-formation system" (Gopnik, 2000). In this paper, she argues for an evolutionary argument to explain why, for humans, the process of understanding something can be inherently pleasurable.

In constructing her argument, she suggests that a parallel can be found in the argument that evolutionary biologists and psychologists have made for the qualitative experience of pleasure, particularly in the evolution of sex. The practical (instrumental/functional) reason for sex is to reproduce in order to ensure the survival of the species. However, individual organisms, within a species, do not have any sense of these evolutionary plans. The essential function of the gene is to reproduce, as the survival of the species depends on it. So evolution, in its blind yet creative manner, has come up with a perfect solution—providing a local incentive that will lead to meeting the global goal. It has done so by making sex inherently fun, i.e., individuals within a species indulge in sex not to propagate the species but because they enjoy it.

Building on this, Gopnik argues that for a species that depends on higher-order thinking to make sense of the world, namely, humans, it makes a lot of sense for evolution to make the process of higher-order thinking itself fun, pleasurable, or fulfilling. Completing the analogy, Gopnik writes:

> … explanation is to theory-formation as orgasm is to reproduction. It is the phenomenological mark of the fulfilment of an evolutionarily determined drive. From our phenomenological point of view, it may seem to us that we construct and use theories in order to achieve explanation or have sex in order to achieve orgasm. From an evolutionary point of view, however, the relation is reversed, we experience orgasms and explanations to ensure that we make babies and theories. (p. 300)

Pleasure is the incentive that evolution has provided to make us continue to think and understand. The physicist and educator Frank Oppenheimer made a similar argument when he suggested that "Understanding is a lot like sex; it's got a practical purpose, but that's not why people do it normally." Or as Gopnik (2000) suggests, "finding an explanation for something is accompanied by a satisfaction that goes beyond the merely cognitive" (p. 311).

We suggest that we should learn from the blind intelligence of evolution and focus also on the proximal motivators for learning in the STEM disciplines, i.e., the aesthetic. The aesthetic exists in the pleasure of understanding and figuring things out. It lives in the thrill of the chase and discovery. It appeals in the sense of awe we feel when we confront at the beauty of nature and the immensities of the universe. It endures in the elegance of a proof or in a subtle line of code.

Where the aesthetic does *not* exist is in the approach of most traditional subject curricula and lessons we have in schools and standards-based learning today. There are encouraging signs in some of the work being done in the STEM disciplines, such as the maker movement and discovery science projects, but those are often exceptions rather than the rule. And more importantly, even in learning contexts where the aesthetic is seen, it is ad hoc, with little or no systematicity to how it is introduced and lacking a framework to articulate its inclusion. The aesthetic emerges, more often than not, as a side effect of some other intervention, rather than a goal in and of itself.

In contrast, we suggest, paraphrasing Frank Oppenheimer, that STEM has an *instrumental* purpose, but that is not generally why people would want to do it. We argue that the basis of aesthetics is in the personal and the subjective, in the powerful human impulses of inquiry, communication, construction, and expression (Dewey, 1943).

The question then becomes, what would a rhetoric of aesthetics for STEAM pedagogy look like? How do we find a pedagogical balance to capture both the aesthetic and the instrumental aspects of doing science? In this chapter, we attempt to use our explorations of aesthetics in science, math, and engineering to push beyond an instrumental STEM to a more inspiring STEAM, with practical methods to integrate an aesthetic and affective rhetoric in a science classroom.

Framing a Rhetoric of Aesthetics: Theory and Research

Thomas Conley (1990) defined *rhetoric* as the art of conducting a discourse of persuasion and motivation, depending on context. In our context of STEM teaching and learning, a discourse of persuasion and motivation to encourage science, mathematics, and engineering would require carefully designed pedagogical moves that draw upon the human impulses toward exploration and understanding. It would require us to create opportunities to inspire affective and emotional elements of beauty, curiosity, wonder, and awe. In addition, the design of a rhetoric requires us to understand the frames that would constitute the discourse. Frames, according to Davis and Russ (2015), are "a set of simple elements that organize the perception of a given situation. Framing is how those elements tune the interpretation of a phenomenon" (p. 223).

7 Developing a Rhetoric of Aesthetics: The (Often) Forgotten Link Between Art...

To develop a rhetoric of aesthetics, and in order to frame it properly, we build on the work of previous scholars and researchers. For instance, Dewey in his book, *Art as Experience*, makes a case for a transactional framing of the aesthetic experience. He wrote:

> In order to understand the esthetic in its ultimate and approved forms, one must begin with it in the raw; in the events and scenes that hold the attentive eye and ear of man, arousing his interests and affording him enjoyment as he looks and listens... [One] should be carried forward, not merely or chiefly by the mechanical impulse of curiosity, not by a restless desire to arrive at the final solution, but by the pleasurable activity of the journey itself. [sic] (Dewey, 1934/2005, pp. 2–3)

Dewey suggests here that understanding should be seen as a "pleasurable" *experience* of doing science and math, where the idea of aesthetic experience is one of unfolding over time. He makes an argument for the value and pleasure of engaging with the process of science, not just the spark of curiosity or satisfaction of the solution. Instead he speaks to *the trajectory of engaging* with process—with a journey that is more than just the destination, with not only the question or the answer but everything that lies in between them.

This is a transactional framing of the aesthetic experience, where people and their worlds mutually interact and co-create meaning (Dewey, 1934). So, Dewey argued that a learning experience is somewhat unique to the individual learner and must be seen as a form of unfolding interaction over time, which can also be collectively shared and understood. In this view, the aesthetic experience is characterized by a sense of heightened engagement, as well as a feeling of anticipation, akin to watching a thoughtfully created book or movie. The pieces work together holistically, and it is through this process of engagement that we create experiences that can be truly transformational. Clearly, this idea of experience is not a static one but rather speaks to a dynamic, dialogic process of interaction between the learner and their world (a world where most STEM experiences happen in contexts mediated by teachers, textbooks, schools, and curricula). Through this process learners shape and construct their understandings and meanings, over time. Disciplinary frames play a crucial role in this process as do certain broad aesthetic themes and ideas.

In developing our approach, we were also guided and inspired by theoretical and empirical work of Girod and Wong (2002), Girod (2007), Pugh and Girod (2007), and Jakobson and Wickman (2008). In their work, these authors offer a definition of aesthetics as a mélange of emotional responses elicited from a variety of experiences (Dewey, 1934/2005). Jakobson and Wickman (2008) connected the positive rhetoric of aesthetic connections in science to what students pay attention to or ignore. Studying student and teacher's scientific discourse, they identified that aesthetic connections shaped students' lived experiences and transformed science content for them.

In particular, our work is deeply connected to and builds on Girod's (2001) study of teaching in a science classroom. In his study, Girod compared two fifth-grade classrooms, one interlaced with a rhetoric of aesthetic connections in science and the other focused on conceptual understanding. Using the quantitative comparisons of students' feelings toward science and conceptual understanding (measured across several different points in time), Girod's analysis demonstrated that, when presented with science in aesthetic terms, students learned more, had deeper understanding of the concepts, and forgot less than students in the control classroom. As he wrote:

> Teaching for aesthetic understanding brings students to high levels of conceptual understanding while simultaneously bolstering more positive feelings toward science and fostering changed action and renewed interest in exploring and engaging with the world. (p. 229)

These studies, though few in number, suggest that a curriculum that includes the aesthetic (or one that places it at the center of the learning experience) can have significant positive impact on student engagement and learning.

In addition, our work builds on theoretical work by Girod (2007) where he offers four themes for considering the aesthetic in science curriculum design. These four themes are listed and briefly described below:

1. *Beauty in experience*: The idea that beauty in science lies in the nature of the experience of doing science as scientists engage in scientific research and creativity
2. *Intellectual beauty*: Beauty in the representation of scientific ideas, in the simplicity and elegance of ideas that explain the most complex of phenomena with ease and grace
3. *Sublime*: The feeling of reverence, wonder, and awe toward the power and grandeur of nature
4. *Beauty as truth*: The beauty in recognizing the truth and the fundamental structures that govern the universe, including a sense that science reveals the grand design of the world

Girod's four themes for considering the aesthetic resonate deeply in our work. It is these themes that we utilized in three different studies, described below. Along the way, however, we also came to develop a new threefold approach toward the aesthetic in STEM that, perhaps unsurprisingly, has a strong affinity with Girod's work while extending it further.

The three studies we describe were meant to help us understand the aesthetic framing of science in three different discursive contexts. Briefly, the first study focuses on the rhetoric of science popularization, the second on the personal experiences of some of the world's top astrophysicists, and the third on comparing the results on memory and experience of aesthetic and instrumental framing of scientific texts. Each of these studies allows us to both build on Girod's themes and to extend their reach, allowing us to revisit his work to refine and develop it further. We offer our new framework at the end of the three studies noted below and touch upon how this new rhetoric of aesthetics has been instantiated in a STEM-related teacher professional development curriculum.

Study 1: A Rhetoric of Aesthetics in Popular Science: The Case of Cosmos In a qualitative analysis of representation of science in popular culture, we thematically analyzed the scientific discourse in a popular and critically acclaimed television documentary series, *Cosmos: A Spacetime Odyssey*. Essentially, we analyzed the transcripts of all 13 episodes of the series, beginning by examining all the videos to

7 Developing a Rhetoric of Aesthetics: The (Often) Forgotten Link Between Art...

verify accuracy of the transcripts and to familiarize ourselves with the data (Bazeley, 2013). We also used this viewing to highlight appropriate positions for further detailed analysis in the transcripts.

Then, we coded the transcripts using a qualitative coding software HyperRESEARCH, using Girod's (2007) themes as a frame of reference for identifying initial codes, which could then be challenged, teased apart, or revised as and when new themes emerged. We employed a bottom-up approach that allowed the coding of emergent themes (DeCuir-Gunby, Marshall, & Muculloch, 2011) focused on a rhetoric of aesthetics—to present an appealing pedagogical approach to science. Multiple iterations of coding helped ensure that all the instances and examples of themes/codes in the text were thoroughly identified (Anfara, Brown, & Mangione, 2002).

Our analysis of the transcripts identified five frames, four of which mapped on to Girod's themes, with one additional emergent frame. Out of Girod's four themes, we found the concept of *beauty in sublime* to be a prominent frame. Here is an example where Cosmos portrayed science as sublime, capable of inspiring awe and wonder and even fear:

> How can we humans, who rarely live more than a century, hope to grasp the vast expanse of time that is the history of the cosmos?... In order to imagine all of cosmic time, let's compress it into a single calendar year. The cosmic calendar begins on January 1st with the birth of our universe. It contains everything that's happened since then, up to now, which on this calendar is midnight December 31st. On this scale, every month represents about a billion years... In the vast ocean of time that this calendar represents, we humans only evolved within the last hour of the last day of the cosmic year. 11:59 and 46 seconds. All of recorded history occupies only the last 14 seconds, and every person you've ever heard of lived somewhere in there. (Cosmos, Episode 1: Standing Up in the Milky Way)

The second most prominent frame in the show was of *beauty in the representation of scientific ideas* specifically the simplicity and elegance of ideas that explain the most complex of phenomenon with ease and grace. These representations of intellectual beauty were prominent throughout the series.

The third frame, which Girod found to be key to creating a rhetoric of aesthetics in science, was of *beauty in grand design*. Cosmos portrayed the beauty in the grand design and the structure of the cosmos by explaining the patterns that constitute it and dissecting any misconceptions that build to further misconceptions. In this process, the series revealed the beauty inherent in nature and discovery of truth. The use of aesthetic examples ranged widely—from the intricate double helix structure of DNA to evolution and natural selection, to a single theory of gravity that makes heavenly objects dance in elliptical orbits, to the intricate marriage of structure and function of a mere dandelion. Across examples, the series depicted the notion that beauty lies in recognizing the truth and the fundamental structures governing the universe.

The fourth frame of *beauty in experience* itself, as Girod (2007) also found, lies in the nature of the experiences themselves as scientists engage in scientific research and creativity (p. 41). Not only did the show reflect the idea of beauty in discovery and invention, it also reflected the sheer joy in research that most scientists feel.

Examples of beauty in experimenting, thinking about science, and experiencing science were prominent throughout the series.

But, we found that Girod's example offered just one aspect under aesthetic experience. Another aspect of aesthetic experience occurred at a somewhat meta-level where the design of the show itself attempted to create an aesthetic experience for the viewer. Cosmos used visuals and verbal cues to not only hook its viewers but to create an aesthetically satisfying experience. The choice of music (typically from the western classical canon) used in the show attempted to construct an experience for the viewer of going on an adventurous, beautiful, exciting, or uplifting pleasing journey. For instance, the evolution of life on earth was orchestrated by the host of the show with Maurice Ravel's Bolero playing in the background. The show used the Ship of the Imagination as a tool to help viewers experience things that would otherwise be impossible, such as traveling inside the body. It asked questions as baits, to compel viewers to think about a recently introduced scientific concept.

The fifth, and emergent, frame that we found highlighted the representation of the role of scientists as being adventurers, detectives, and explorers. Consider the two examples below in this frame:

> But for one man, Copernicus didn't go far enough. His name was Giordano Bruno, and he was a natural-born rebel. He longed to bust out of that cramped little universe. Even as a young Dominican monk in Naples, he was a misfit. This was a time when there was no freedom of thought in Italy. But Bruno hungered to know everything about God's creation. (Cosmos, Episode 1: Standing Up in the Milky Way)
>
> Science works on the frontier between knowledge and ignorance. We're not afraid to admit what we don't know. There's no shame in that. The only shame is to pretend that we have all the answers. Maybe someone watching this will be the first to solve the mystery of how life on Earth began. (Cosmos, Episode 2: Some of the Things That Molecules Do)

The emphasis of the show on the beauty in science and the excitement of the profession captures the essence of Carl Sagan's "Cosmos perspective" and aligns well with the aesthetic framework described in this paper. It suggests a shifting of focus from instrumental reasons for learning science to ones that connect with deeper themes of aesthetic experience—making a stronger case for STEAM in seeing it as the inclusion of a rhetoric of the aesthetic in the teaching and learning of STEM disciplines.

Study 2: A Rhetoric of Aesthetics in Personal Scientific Narratives: Listening to Cosmologists The second study was a qualitative analysis we conducted of 27 in-depth existing interviews with top-ranked cosmologists to better understand their personal rhetoric of science. The idea behind understanding their personal lens for science was to look for differences in the types of instrumental and aesthetic rationales they considered for and within their personal work. The completed interview transcripts from the 27 cosmologists done by Alan Lightman provide the content of Lightman and Brawer's (1990) book *Origins: The Lives and Worlds of Modern Cosmologists* the data for our exploration. We again started with Girod's themes for aesthetic understanding and our categories for instrumental motivations. We completed a preliminary coding of these transcripts in the qualitative software, NVivo. We then developed codes to reflect beauty in the sublime, intellectual beauty,

beauty in grand design, and beauty in experience. Within each of these, we identified and analyzed emergent subcategories to incorporate into the framework.

The interviews covered a range of technical and personal detail, offering insight into the motivations these scientists had for entering the field and the affective nature of their continued participation and motivation toward STEM fields. The interviewees discussed books that piqued their interest and inspired further investigation by using aesthetic terms (e.g., being "turned on" to science, developing "serious interest"). This theme of emotional arousal consistently grew as the astrophysicists described their career trajectory, even if their involvement in science had included instrumental or practical reasons. Such sustained aesthetic references spoke to the compelling nature of aesthetic appreciation that Girod and Wong (2002) identify as being a key characteristic differentiating aesthetic understanding from conceptual understanding.

One of the codes that appeared most frequently is that of *beauty in experience*. The scientists often discussed their research experience in affective terms, talking about their work and their reactions to others in terms of *feeling* "worried," finding things "fun" or "exciting," or being "bothered" by certain ideas. This resonates with Dewey's idea of the heightened emotional or affective dimensions of any aesthetic educative experience (Dewey, 1934). It is one kind of strong aesthetic connection to the material, which drives continued engagement with problems and a passion extending that engagement. For instance, Robert Wagoner, astrophysicist at Stanford, in attempting to communicate the importance of his work explained, "I really got worried about people being too concerned with their everyday life and not looking out to be aware of their cosmic environment, to put things in perspective."

This attempt to evoke emotional reactions to the vastness of the cosmos is an almost exact definition of our third category: *beauty in sublime*–or beauty in awe and wonder. This reiterates the importance that an aesthetic understanding of science can hold as a means of communicating value and providing accessible points of contact with the public.

Overall, what we see when we look at the personal rhetoric of these astrophysicists is that successful STEM practitioners do not focus on the instrumental aspects of their field when speaking about what is exciting or motivating or what drives curiosity in science. Instead, they focus on the highly aesthetic and affective dimensions of participating in STEM fields. And this incorporation of the aesthetic drives a more STEAM-based sensibility toward STEM.

Study 3: The Rhetoric of Aesthetics (and the Instrumental) in Science Texts— An Experimental Test The third study we conducted focused on disentangling the influence of aesthetic and instrumental perspectives on how undergraduate students read and interpret scientific texts. This experimental study was designed along the lines of past cognitive interventions that have investigated the effect of "framing" on memory, recall, engagement, and understanding (e.g., Anderson & Pichert, 1978). In this context, participants were given either an instrumental frame or an aesthetic frame prior to reading a scientific text. Participants in the aesthetic condition were given a framing passage that suggested aesthetic reasons like curiosity, transformative experiences, and intellectual beauty as reasons to study science. Participants in the instrumental frame received a frame that emphasized the value of doing science in terms of economic mobility, enhancing representation, and scientific literacy.

After reading their frames, participants "wrote back" what they could remember about the frames to help them internalize the ideas about what some core reasons were for doing science. Each group then reads their first scientific text passage (about virus reproduction), completed a distractor activity, and wrote back all they could recall about the passage. Following this they were given a prompt to write about the reasons they were given for studying science (the frame), then read the second scientific text passage (about the structure of atoms), completed another distractor activity, and then wrote their final recall.

To determine whether the framing passage affected participants' recall of scientific texts, identifiably unique statements made in each scientific passage were given a code and compared to participants' responses. The results of the study were inconclusive. It is true that participants in the aesthetic framing group did recall a higher number of ideas from both scientific text passages than their instrumental counterparts, yet (possibly due to a small sample size) it was not at a statistically significant level.

The next step of the analysis focused on whether there was a difference in *which* statements participants recalled. To determine whether the frame affected *the way* participants constructed their recalls, all recall responses were run through the Linguistic Inquiry and Word Count (LIWC) database and analyzed along 28 dimensions. There was only one significant difference between the groups, but again looking at effect size revealed additional differences between the constructions of the recalls that suggest different approaches and interpretations of the scientific texts.

One of the complications in the study was a possible interaction with the content of the scientific passages. Participants, in reporting their enjoyment of reading the passages, revealed a statistically significant preference within the aesthetic group for the virus passage over the atom's passage along four dimensions. In contrast, the instrumental group preferred the same passage but only along two dimensions. This offers tentative support to the hypothesis that participants who received an aesthetic framing passage not only had stronger emotional reactions when considering the passages (in their more clearly defined preference for the virus scientific passage) but also enjoyed the process more than their counterparts in the instrumental framing group.

The results of this experiment offer preliminary support for our hypothesis that the framing of the participants' perspectives (along aesthetic or instrumental dimensions) influenced their recall of scientific text and the affective quality of their experience with it. This study offers an exploratory first step toward our understanding of how those perspectives may influence what students pay attention to and their enjoyment of the process. Additionally, these findings demonstrate that highlighting aesthetic reasons for pursuing STEM knowledge at the very least did not hamper participants' performance and arguably increased their enjoyment of the process as compared to their peers who had been given traditional, instrumental reasons to study science. This suggested that it is possible that even those with instrumental motivations to push participation in STEM might benefit from including an aesthetic perspective in their campaigns and curriculum.

A Rhetoric of Aesthetics for STEAM: Three Fractal Frames

In some form, the three studies described above speak to the importance of aesthetics in thinking and learning in the STEM disciplines. Moreover, it is a specific form of aesthetics that we are describing here; it is in an aesthetic experience in the Deweyan sense of the word. It is a perspective that sees deep commonalities between the artistic and the scientific. Describing the artist experience, Deweyan scholar Philip Jackson (1998) wrote:

> Our interactions with art objects epitomize what it means to undergo an experience, a term with a very special meaning for Dewey. The arts do more than provide us with fleeting moments of elation and delight. They expand our horizons. They contribute meaning and value to future experience. They modify our ways of perceiving the world, thus leaving us and the world itself irrevocably changed. (p. 33)

In the quote above, the phrase "interactions with art objects," according to the perspective we are constructing here, could be replaced by the phrase "engagement with STEM." We suggest that the STEM fields play the same role in our consciousness that the arts do; in that, as Dewey (1934) or Jackson (1998) suggests, they expand our vision of knowledge and learning and develop a wider perception and experience of the world.

It is this unity of the experiences that the arts and STEM disciplines create (in terms of Dewey's aesthetics) that we believe is the way in which STEM and STEAM are deeply and powerfully interconnected—where the aesthetic is the missing link between STEM and STEAM. The addition of the "A" to STEM is not just the addition of the arts to the curriculum. Instead, it is an awareness that the arts and the sciences are more deeply connected than traditional disciplinary boundaries reflect. They are connected in ways that integrate them as inherently humanistic enterprises which allow us to experience and engage with the world in profound and transformational ways.

We seek to coalesce these disparate themes (Dewey's ideas from *Art as Experience*, Girod's research on themes of beauty in science, and our own work described above) into a rhetoric of aesthetics, intersecting around STEAM learning. Though the three studies were broadly based on Girod's themes, we also over time began to develop our own framework to support a rhetoric. Girod's work, though important, offers these four themes, but provides little structure for how the themes flow into one another, and their interrelationships. In contrast, our rhetoric, described in greater detail below, seeks to capture the entire cycle of engaging in STEM practices: from curiosity to the process of seeking answers, to a sense of completion, that in turn leads to new curiosities to explore (see Fig. 7.1). This forms a

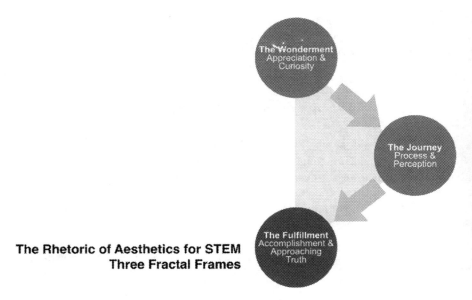

The Rhetoric of Aesthetics for STEM Three Fractal Frames

Fig. 7.1 The three frames that make the rhetoric of aesthetics

powerful virtuous cycle that seeks to maintain the same threefold sequence at different levels of learning—from the beginner to the professional scientist, mathematician, or engineer. We see this as a recurring fractal pattern, each phase informing the other. We call the three frames *The Wonderment*, *The Journey*, and *The Fulfillment* (see Table 7.1). We define and discuss each of these frames below and further define the subcategories that constitute each frame, in an aim to describe a rich picture of all that goes into an aesthetically driven vision of STEM.

1. **The Wonderment.** The first frame of aesthetics is that of the affective reaction—a sense of wonderment. It is the beginning of the aesthetic experience, building a sense of anticipation for future engagement. We see it has having two key subcategories: a sense of *appreciation* and a sense of *curiosity*. Both of these can vary a great deal depending on the knowledge of the individual. For instance, the appreciation of science or mathematics or engineering of a beginner would be very different from that of an expert.

    1.1 *Appreciation*. A cognitive-emotional reaction of awe, admiration, or respect inspired by feelings of astonishment, a sense of the sublime, or fear in nature and its understanding/explanation. In the beginning, appreciation may be for nature and the world, and as we learn more, our appreciation migrates toward the more abstract intellectual beauty and power of ideas and representations. Abstract concepts that help understand nature, in themselves, can inspire an aesthetic or affective reaction. This involves representations of the workings of nature, or explanations of its complexities, where one arrives at a theory and can finally exclaim admiration and appreciation. For instance,

7 Developing a Rhetoric of Aesthetics: The (Often) Forgotten Link Between Art... 131

Table 7.1 The three frames and their subcategories

#	Frames		Subcategories
1	The Wonderment	1.2	Appreciation
		1.1	Curiosity
2	The Journey	2.1	Process
		2.2	Perception
3	The Fulfillment	3.1	Accomplishment
		3.2	Approaching truth

Einstein's theory of special relativity leading to the iconic equation, $E=mc^2$, and Euler's formula, $e^{\pi i} + 1 = 0$, which has been called the most beautiful equation in mathematics, are some examples of appreciation that celebrate intellectual beauty. Laypeople and scientists alike can appreciate the beauty of many scientific phenomena, like a sunset, but scientific and mathematical knowledge can further lead to an appreciation of beauty in increasingly abstract ways (Girod, 2007). Appreciation can also lead to a sense of curiosity, a drive to learn more about something you appreciate, thereby paving the way for more scientific and mathematical knowledge.

1.2 *Curiosity.* A cognitive-emotional desire to seek, to anticipate, and to understand and/or solve problems or phenomena. This is the intellectual equivalent of an itch that must be scratched. The desire to learn about the unknown is, arguably, a fundamental human trait. We are capable of reacting to feelings of awe, admiration, and respect with sense of curiosity that kindles a desire to seek, anticipate, and solve problems and answer questions, in essence, to understand. Such moments of curiosity and anticipation are at the precipice of participation in STEM. Reacting to nature, one may feel like a detective who wants to solve new mysteries. In engineering, it may be the desire to tinker and play to construct new artifacts. This anticipation further fuels, organizes, and develops educative experiences (Girod & Wong, 2002). Teachers can tap into this feeling of curiosity and anticipation by creating experiences that inspire such emotional and cognitive responses.

2. **The Journey.** This is the *process* part of the STEM experience. It varies across the STEM disciplines, but across all are similarities in (a) learning the nature of the discipline, namely, norms, methods, knowledge, and purposes of discipline (Gardner & Boix-Mansilla, 1999) and (b) a process of socialization into the discipline, though again this has an important individual construct. This stage builds upon the Deweyan idea that "having an experience" can transform one's understanding of oneself and the world. As we have noted, Dewey (1934) viewed an educative experience as not just a question or a solution but a complex process—a journey. This second frame of the journey can be divided in two subcategories: the *process* and the *perception*.

2.1 *Process.* This is the experience of participating in STEM, where open scientific and mathematical inquiry itself can naturally turn to an aesthetic experience. Participating in the STEM-based inquiry, a learner works like a detective, aiming to solve the mysteries of the world using a sense of wonder as fuel and guidance. The act of exploration is, in itself, transformational. This can be an individual or group process that includes within it the act of finding and defining a problem and developing personal/shared strategies to solve it, leading to a shared sense of values and criteria for what is a good or bad solution. Throughout the journey, the process of going through the adventure becomes an educative experience, inspiring a deeper understanding of the world. This turns naturally into shaping how we look at the world—changing perceptions.

2.2 *Perception.* This is the transformative experience of looking at the world while doing or having participated in explorations in STEM. Richard Feynman illustrates this in his story about a conversation he had with a friend who claimed looking at a flower as a scientist diminished experiencing its beauty:

> …I see much more about the flower than he sees. I could imagine the cells in there, the complicated actions inside, which also have a beauty…the science knowledge only adds to the excitement, the mystery and the awe of a flower. (as quoted in Sykes, 1981)

Feynman demonstrates that his understanding and appreciation of the world is fundamentally changed by his knowledge about biology, which in turn changes *him* and the way in which he interacts with nature. Even if one has not achieved a solution to their question, or does not fully understand how the nature works, their way of looking at the world has changed because of their participation in STEM.

3. **The Fulfillment.** The third frame is the feeling of fulfillment. It is the (possibly temporary) culmination of engaging in the STEM-related activity, temporary only because it often leads to new wonderments (appreciation and curiosity) and the start of a new journey (process and perception). The subcategories here are a sense of *accomplishment* and the discovery of *truth.*

3.1 *Accomplishment.* The feeling of pride at having reached the end of a curious exploration. For example, this may involve finding of an explanation or an answer for a problem/question and is often characterized by a sense of completion tempered with the knowledge that more phenomena remain to be understood and more problems remain to be solved. The sense of accomplishment can be seen as small feelings of fulfillment that emerge as a result of having reached a conclusive explanation of a curiosity. A curiosity in STEM might be big or small depending on the task involved and the intricacies of the journey it entails. However, all curiosities, big and small, eventually reach an end that gives a sense of having accomplished something. This sense may not be as powerful as an overarching sense of fulfillment but contributes toward it in small wins. For example, spending months figuring out a way to successfully land a rover on Mars does not have to end in a discovery of truth but gives a sense of having achieved a breakthrough solution that adds to the overarching sense of fulfillment.

3.2 *Approaching Truth.* This is the sensation of having genuinely engaged with the world and understood it on its own terms. It is a sense that we have conducted a reality check that the abstractions and constructions created help explain, exploit, and predict patterns and rules in the real world. It is a sense of accessing truths, of grasping how the world works, independent of us, however fleeting that feeling may be. It is the sense of getting an inside glimpse at the "grand design." There is sometimes the feeling of a spiritual, religious, or even mystical quality to this component of an aesthetic experience (see Fig. 7.2).

The Three Fractal Frames A rhetoric of aesthetics in STEM is made of these three key frames that we see as pieces of a recurring, recursive, spiral (akin to a fractal pattern), feeding each other through a cognitive-emotional transaction. Wonderment is not simply one beginning, just as fulfillment is not the end. Fulfillment leads to new appreciations, questions, and wonderment—leading to new processes and perceptions. The cycle has the potential to go on forever—deeper and deeper. This is what we mean by *fractal frames*; these three frames and their relationships continue as long as we have the energy to pursue these ideas.

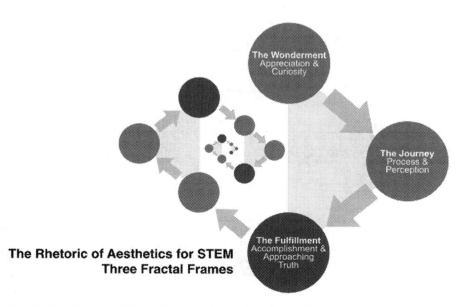

Fig. 7.2 The three fractal frames that make the rhetoric of aesthetics

Lessons for Practice: Designing a Rhetoric of Aesthetics in STEM

Moving from simply a rhetoric of aesthetics to its actual instantiation in curriculum or classroom practices is not a trivial task. Some might even consider it a wicked problem (Koehler & Mishra, 2008; Rittel & Webber, 1973). There are some particular challenges in developing a curriculum that values the aesthetic, primarily because it encourages an experiential nature that differs from traditional or common schooling approaches but also because it evolves over time. This requires that we move beyond thinking about content to considering a host of intangible elements that go into the experience that teachers of such a STEAM curriculum would face. In the section below, we articulate our approach broadly and then discuss specifics of how the revised threefold rhetoric we propose above plays out in the specifics of curricular enactment.

Exemplifying a STEAM-Based Aesthetic Framework Within Teacher Professional Development We will contextualize our discussion within a specific STEM-focused teacher professional program (the MSUrbanSTEM program; msurbanstem.org), which was a yearlong blended professional development fellowship experience for STEM teachers in Chicago Public Schools. Two authors of this chapter have worked closely with 124 in-service Chicago teachers/fellows, across 3 different cohorts spanning 3 years. The yearlong fellowship program focused on STEM and leadership, using John Dewey's philosophy of learning by doing and experiencing, along with Mishra and Koehler's (2006) TPACK-based approach to repurposing technology in practice. One of the many purposes of the fellowship program was to provide teachers with approaches to take leadership in their individual contexts, to find compromises and strategies for implementation between their instrumental curricula of STEM and a rhetoric of aesthetics.

The instructional team worked closely with the teachers/fellows, both face to face and online, throughout the yearlong fellowship. The program began with 11 days of face-to-face work in summer and was followed by online components that spanned the school year, with 4 full-day Saturday meetings (two in each semester). Fellows worked on a range of projects during the year, what we describe as micro-design projects, macro-design projects, and reflections on the total package of implementing both TPACK and aesthetic understandings (Koehler et al., 2011). The rhetoric of aesthetics played a key role in the design and enactment of the curriculum (as described in greater detail below), in an attempt to bring more STEAM to these teachers' STEM.

In considering the rhetoric of aesthetics and its role in curriculum development, we must remember what Dewey said about experiencce when he wrote that "The belief that all genuine education comes about through experience does not mean that all experiences are genuinely or equally educative" (Dewey, 1938, p. 25). An experience is a transaction that happens between an individual and the variable of elements of their environment and the world around them. It is not just a psychological phenomenon that happens exclusively "within" us but rather that experience takes place in the world itself. It is made up of our continuous interaction and participation with objects, situations, and events that constitute our environment.

As we consider the role of the aesthetic in teaching and learning, one of the first things to consider is the fragility of the aesthetic encounter itself, which means that the nature of the aesthetic is such that it requires elements to work together in a holistic, synergistic, and coherent manner, as it is easy for the intended experience to fall apart. Thus we, in the MSUrbanSTEM project, took great care to consider (in so far as possible) every aspect of the instructional system, to create an environment where risk-taking and failure were acceptable, where wonder and new ways of looking at the world were emphasized, and where beauty and truth (in the STEM disciplines) were valued and celebrated.

This meant the MSUrbanSTEM program needed to have certain characteristics that would allow the rhetoric of aesthetics to play out successfully in an interactional and transactional way for the teachers. These characteristics included consideration of the following factors: the curriculum or the program had to be *unified*, in that the coursework was integrated and coherent, and strongly connected to practice. It had to emphasize *combinatorial creativity* (or a view that creativity emerges when different elements or disciplines combine in unique ways), through a focus on repurposing technology and using new lenses on the curriculum. It was built on the idea of *deep play*, which is an active, collaborative engagement with rich problems of practice and a focus on finding new ways of seeing the world and new approaches to teaching. Finally, and most importantly, the curriculum and its instantiation in the classroom context needed to have a *dramatic quality*, or a compelling narrative driven by wonder and curiosity and aesthetics qualities we have described above.

In the section below, we describe, briefly, how each of these elements played out through the yearlong program. It is important to point out that though we tease apart these elements in our writing, it is much harder to do so in practice. In practice these elements are interconnected and work together to create a coherent whole, which is often greater than the sum of its parts. As this is only one chapter, we cannot include a complete description of how the aesthetic played out in the entirety of the program. But to exemplify these ideas to some degree, we take the fractal frames informing our rhetoric of aesthetics and provide examples of how each of them is intentionally integrated within the MSUrbanSTEM program.

The Wonderment (Appreciation and Curiosity) One of the tasks done at the beginning of each day is what we call the sharing of a World of Wonder moment, i.e., WOW! moment. These WOW! moments were introduced by the instructors and consisted of sharing something in the world around them that intrigued them or made them wonder.

For instance, one teacher/fellow shared and discussed a similarity she saw in patterns of bubbles in a bowl of lentil soup boiled on a gas stove and a specific atmospheric pattern known as "cloud trains." Both these phenomena had been independently observed by one of the instructors (and authors of this chapter). Upon digging deeper, the similarity between these two phenomena was revealed.

The teachers/fellows were promoted to do this at the start of every day, in order to encourage the practice of wonder in themselves (ideally leading to an instantiation of it in their teacher practices). Though this began somewhat slowly, as the days went on, the fellows began to share a wide range of examples that they had been observed in their own lives: from how traffic is controlled in Chicago's express lanes to how the new soda-creation machines actually work and from how many more steps a shorter person will take in their lifetime compared to someone taller to wondering whether people are differentially attractive to mosquitoes.

As these examples indicate, the kinds of problems that arose were diverse and deeply personal (for instance, the one about number of steps taken in a lifetime was suggested by a teacher who was herself quite short and self-confessedly had to almost run to keep up with others even while walking). Even more interesting than how they practiced the activity in the professional development sessions was how the fellows took up and integrated this practice into their own classrooms. Some of our teachers/fellows used our activity "as is" with their science students, while others created variations for their unique contexts. For example, one teacher/fellow created a "Wall of Wonder" in their classrooms, which was a bulletin board where students could post questions that intrigued them—they would then collectively choose to investigate them further. Essentially, this activity promoted the fellows to be more present in the world to view it from an interdisciplinary-aesthetic STEAM sense of science and wonder. Through this they could see the world as a source of questions and mystery (the emotional component) *and*, at the same time, amenable to analysis and understanding, often through a disciplinary lens (the cognitive component).

The Journey (Process and Perception) One could argue that the WOW! activity above was key to learning and investigation in STEM. That said, the fellowship provided teachers multiple opportunities to engage in the inherent pleasure in the experience of *doing* STEM in an aesthetic, real-world, interdisciplinary way. This included such activities as visits to Chicago's Museum of Science and Industry where the teachers/fellows were prompted to look at the exhibits through the lens of their disciplinary interests or having them engage in a mini-maker fair, where they constructed a range of artifacts using everything from maker kits, mini-circuits, straws, and Play-Doh. The idea of "engineering" a solution to a problem was actively promoted in multiple ways. For instance, the teachers/fellows created musical instruments using whatever real-world objects they could find around them. In each case, the idea was to engage our teachers/fellows in authentic STEAM-related experiences to explore the aesthetic pleasure and beauty of *doing* work in the sciences. Our teachers/fellows took these ideas and implemented them in their classrooms as well, in various ways. Specifically, in their yearlong projects, many of our teachers/fellows took on key aspects of what it meant to do science, mathematics, and engineering; and they had their own students engage in the actual process of *doing* STEM. In that sense, the perception of science shifted from being a narrow, cold, or rigid mechanical task into one of activity, excitement, engagement, and creativity.

The Fulfillment (Accomplishment and Approaching Truth) Achieving a sense of accomplishment, or a feeling of engagement with the world, or approaching some kind of truth, is difficult to achieve in a classroom context. But at least experientially, there were moments in the MSUrbanSTEM program where it appeared that we had genuinely, even if for a moment, gotten there. We provide one such example below.

A few days into the program, one of the instructors (and co-author of this chapter) introduced the statistical concept of the "long tail" in business and education to the students. That led to a discussion of power laws and what those curves mean in the real world. At that point, the idea of Benford's Law came up. Benford's Law, also known as the first-digit law, is a counterintuitive fact that notes how in all kinds of listings of numbers in the real world, the digit 1 tends to occur as the first or leading digit far more frequently than expected. This is true of sets of numbers as disparate as electricity bills, stock prices, lengths of rivers, physical constants, population numbers, and so on. This "strange" result led to a great deal of discussion, until two teachers/fellows took up the challenge of explaining it to the entire class. They approached it from two very different directions—one going with an intuitive sense of numbers and growth and the other using mathematical formulas for power curves and exponential growth. Through that process, the entire class experienced and learned to appreciate the way abstract mathematical ideas can take what initially appear to be anomalous phenomena in the world and make sense of them through the language of mathematics, with relatively intuitive and clear outcomes of mathematical principles. Here, there was a sense of accomplishment in grasping some fundamental truth about the world and a sense that the world had opened itself up for investigation and revealed its "grand design."

Conclusion

In this piece, we have aimed to promote a model of learning within the STEM disciplines that is predicated upon aesthetic ways of knowing, thinking, and exploring the world. We suggest that bringing such an aesthetic sensibility into the arena of STEM provides us with a lens for STEAM, which spans the arts, sciences, and many other disciplines. Our view of STEAM, in this way, allows the value of interdisciplinary learning to emerge in contexts that value the very impulses that make us human—including curiosity, excitement, inspiration, exploration of ideas, testing and probing the world, and appreciating beauty (in the world and in the explanatory structures we create).

We have aimed to construct our argument along several lines of theory, drawing on classic educational philosophers such as Dewey (1934), to more recent works that bridge education and aesthetics (Girod, 2007), to principles of human psychology (Gopnik, 2000), and to work that explores the history of science and technology or accounts of STEM practitioners (Chandrasekhar, 1987; Orrell, 2012;

Root-Bernstein & Root-Bernstein, 1999). We have built upon all of these ideas to make a clear case for the value of employing a rhetoric of aesthetics in STEM teaching and learning—to draw us toward a more STEAM-based view of learning.

Moving beyond theoretical exploration, we also have shared several accounts of recent studies from our own lines of inquiry, which help us to think more about how these constructs play out in research. Each of the three studies we described, in connection with our rhetoric of aesthetics, have helped us develop some practical considerations for teaching STEM in a more aesthetic manner. The rhetoric of aesthetics that we have developed out of this work is focused on the threefold fractal frames of *The Wonderment, The Journey,* and *The Fulfillment,* to help us paint a picture of STEAM which reflects how people might feel and experience STEM in ways that inspired practitioners often do.

In the provision of any set of frames or lenses for viewing subject matter, it is important to consider what these mean for teacher-practitioners, since teachers are the mediators of experiences that learners have in schools. We discuss and expand on what this rhetoric has looked like in an example of STEM teacher professional development—showing examples of how these frames The Wonderment, The Journey, and The Fulfillment have been instantiated for teachers. Furthermore, these frames have (at least in our preliminary experiences) helped teachers to rethink and redefine their lenses for STEM—moving them toward a more aesthetically driven STEAM perspective and changing their own view on science to manifest a personal rhetoric of aesthetics in how they look at the world.

The blending of aesthetic ways of knowing with STEM disciplinary teaching and learning was always the driving force of this work—in the research, practice, and theory pieces that we propose here. Along the way, it became clear that this rhetoric of aesthetics for STEM could provide a uniquely appropriate lens for thinking about STEAM. The terminology for STEAM is still relatively new in the arena of educational research and literature. However, the foundations of STEAM can be found in the work of Dewey, who proposed a much more interdisciplinary, experiential, and aesthetically driven view of curricula and student learning.

Our work here draws upon some of the foundational ideas about learning, as well as the accounts and descriptions of how actual STEM experts often experience these disciplines in inspired ways. Such ways of knowing focus on concepts that are as integral to the arts as they are to the sciences, as well as to all disciplines of human knowledge. This reveals a sensibility for STEAM rooted in STEM learning experiences that include more authentic or real-world learning experiences or the blending and blurring of disciplinary lines and which integrate activity, curiosity, and more emotional connection to the complexity of STEM. All of these elements marry well with the ways in which scholars have begun to think about STEAM recently (Bequette & Bequette, 2012; Boy, 2013; Brophy, Klein, Portsmore, & Rogers, 2008; Jolly, 2014, 2016; Piro, 2010; Radziwill, Benton, & Moellers, 2015).

In our proffered frames and rhetoric, there is nothing dramatically new or wildly inconsistent with existing underpinnings of learning and human psychology. But what we do offer is a subtly new interpretation of—and practical or theoretical frame for—understanding STEAM as an aesthetically driven approach to STEM. We

7 Developing a Rhetoric of Aesthetics: The (Often) Forgotten Link Between Art…   139

share a construction and constellation of ideas to help teachers and students consider and explore STEM in ways that uphold their sense of beauty, wonder, awe, curiosity, and the inherent pleasure of figuring things out. Through this, the aim is to give them the opportunity to experience new knowledge much in the way that artists, scientists, mathematicians, and the most inspired thinker across disciplines do. If we need such creative, inspired, and interdisciplinary problem-solvers to become involved in STEM areas, then education must consider an approach that speaks clearly to this need.

References

Anderson, R. C., & Pichert, J. W. (1978). Recall of previously unrecallable information following a shift in perspective. *Journal of Learning and Verbal Behavior, 17*, 1–12.

Anfara, V. A., Brown, K. M., & Mangione, T. L. (2002). Qualitative analysis on stage: Making the research process more public. *Educational Researcher, 31*(7), 28–38. https://doi.org/10.3102/0013189X031007028

Bazeley, P. (2013). *Qualitative data analysis: Practical strategies.* London: Sage.

Bequette, J. W., & Bequette, M. B. (2012). A place for art and design education in the STEM conversation. *Art Education, 65*(2), 40–47.

Boy, G. A. (2013). From STEM to STEAM: toward a human-centered education, creativity & learning thinking. In *Proceedings of the 31st European Conference on Cognitive Ergonomics* (p. 3). ACM.

Brophy, S., Klein, S., Portsmore, M., & Rogers, C. (2008). Advancing engineering education in P-12 classrooms. *Journal of Engineering Education, 97*(3), 369–387.

Chandrasekhar, S. (1987). *Truth and beauty: Aesthetics and motivations in science.* Chicago: University of Chicago Press.

Conley, T. M. (1990). *Rhetoric in the European tradition.* New York: Longman.

Davis, P., & Russ, R. (2015). Dynamic framing in the communication of scientific research: Texts and interaction. *Journal of Research in Science Teaching, 52*(2), 221–252.

DeCuir-Gunby, J. T., Marshall, P. L., & McCulloch, A. W. (2011). Developing and using a codebook for the analysis of interview data: An example from a professional development research project. *Field Methods, 23*(2), 136–155.

Dewey, J. (1934/2005). *Art as experience.* New York: Minton, Balch & Company.

Dewey, J. (1938). *Experience and education.* New York: The MacMillan Company.

Dewey, J. (1943). *The child and the curriculum: The school and society.* Chicago: University of Chicago Press.

Dirac, P. A. M. (1963). The evolution of the physicist's picture of nature. *Scientific American, 208*, 45–53. https://doi.org/10.1038/scientificamerican0563-45

Gardner, H., & Boix-Mansilla, V. (1999). Teaching for understanding in the disciplines–and beyond. In J. Leach & B. Moon (Eds.), *Learners and pedagogy* (pp. 78–88). London: Paul Chapman.

Girod, M. (2001). *Teaching fifth grade science for aesthetic understanding.* Retrieved from ProQuest. (UMI Number: 3009113).

Girod, M. (2007). A conceptual overview of the role of beauty and aesthetics in science and science education. *Studies in Science Education, 43*(1), 38–61. https://doi.org/10.1080/03057260708560226

Girod, M., & Wong, D. (2002). An aesthetic (Deweyan) perspective on science learning: Case studies of three fourth graders. *The Elementary School Journal, 102*(3), 199–124.

Good, J., Keenan, S., & Mishra, P. (2016). Education:=coding+aesthetics; Aesthetic understanding, computer science education, and computational thinking. *Journal of Computers in Mathematics and Science Teaching, 35*(4), 313–318.

Gopnik, A. (2000). Explanation as orgasm and the drive for causal understanding: The evolution, function and phenomenology of the theory-formation system. In F. Keil & R. Wilson (Eds.), *Cognition and explanation* (pp. 299–323). Cambridge, MA: MIT Press.

Hoffmann, R. (1990). Molecular beauty. *The Journal of Aesthetics and Art Criticism, 48*(3), 191. https://doi.org/10.2307/431761

Holton, G. J. (1988). *Thematic origins of scientific thought: Kepler to Einstein* (Rev ed.). Cambridge, MA: Harvard University Press.

Jackson, P. (1998). *John Dewey and the lessons of art.* Yale University Press. Retrieved from http://www.jstor.org/stable/j.ctt32bwqn.

Jakobson, B., & Wickman, P. O. (2008). The roles of aesthetic experience in elementary school science. *Research in Science Education, 38*, 45–65.

Jolly, A., (2014). STEM vs. STEAM: Do the arts belong? *Education week: Teacher.* Retrieved from http://www.edweek.org/tm/articles/2014/11/18/ctq-jolly-stem-vs-steam.html.

Jolly, A. (2016). *STEM by Design: Strategies and activities for grades* (pp. 4–8). Routledge.

Koehler, M. J., & Mishra, P. (2008). Introducing TPACK. In American Association of Colleges for Teacher Education Committee on Innovation and Technology (Ed.), *Handbook of technological pedagogical content knowledge (TPACK) for educators* (pp. 3–29). New York: Routledge.

Koehler, M. J., Mishra, P., Bouck, E. C., DeSchryver, M., Kereluik, K., Shin, T. S., et al. (2011). Deep-play: Developing TPACK for 21st century teachers. *International Journal of Learning Technology, 6*(2), 146–163.

Lightman, A. P., & Brawer, R. (1990). *Origins: the lives and worlds of modern cosmologists.* Cambridge, MA: Harvard University Press.

Mehta, R., Mishra, P., & Henriksen, D. (2016). Creativity in mathematics and beyond – Learning from fields medal winners. *TechTrends, 60*(1), 14–18. https://doi.org/10.1007/s11528-015-0011-6

Mishra, P., Terry, C. A., Henriksen, D., & Deep-Play Research Group. (2013). Square peg, round hole, good engineering. *TechTrends, 57*(2), 22–25.

Mishra, P., & Koehler, M. J. (2006). Technological pedagogical content knowledge: A framework for teacher knowledge. *Teachers College Record, 108*(6), 1017.

National Commission on Excellence in Education. (1983). *A nation at risk: The imperative for educational reform. An open letter to the American people. A report to the nation and the Secretary of Education.* Washington, DC.

Orrell, D. (2012). *Truth or beauty: Science and the quest for order.* New Haven, CT: Yale University Press.

Petroski, H. (2011). *The essential engineer: Why science alone will not solve our global problems.* New York: Vintage.

Piro, J. (2010). Going from STEM to STEAM: The arts have a role in America's future, too. *Education Week, 29*(24), 28–29.

Poincaré, H. (1910). Mathematical creation. *The Monist, 20*(3), 321–335. https://doi.org/10.5840/monist19102037

Pugh, K., & Girod, M. (2007). Science, art, and experience: Constructing a science pedagogy from Dewey's aesthetics. *Journal of Science Teacher Education, 18*, 9–27.

Radziwill, N. M., Benton, M. C., & Moellers, C. (2015). From STEM to STEAM: Reframing what it means to learn. *The STEAM Journal, 2*(1), 3.

Rittel, H. W. J., & Webber, M. M. (1973). Dilemmas in a general theory of planning. *Policy Sciences, 4*(2), 155–169. https://doi.org/10.1007/BF01405730

Root-Bernstein, R., Bernstein, M., & Garnier, H. (1995). Correlations between avocations, scientific style, work habits, and professional impact of scientists. *Creativity Research Journal, 8*(2), 115–137. https://doi.org/10.1207/s15326934crj0802_2

Root-Bernstein, R. S., & Root-Bernstein, M. (1999). *Sparks of genius: The thirteen thinking tools of the world's most creative people*. Boston: Houghton Mifflin Co.

Scruton, R. (1983). *The aesthetic understanding: Essays in the philosophy of art and culture*. New York: Metheun & Co.

Sykes, C. (1981). The pleasure of finding things out. *Motion Picture*. London: BBC 2.

Tauber, A. I. (1997). *The elusive synthesis: Aesthetics and science, Softcover reprint of the original 1st ed. 1997 edition*. Dordrecht, The Netherlands: Springer.

Chapter 8
Moving Toward Transdisciplinary Instruction: A Longitudinal Examination of STEAM Teaching Practices

Cassie F. Quigley, Dani Herro, and Abigail Baker

Introduction

The emergence of STEAM (science, technology, engineering, arts, and mathematics) education, a transdisciplinary approach that focuses on problem-solving (Delaney, 2014), is occurring worldwide. However, there is little available literature regarding the efficacy of STEAM practices. As a result, educators are attempting to implement new teacher practices without a solid conception of how to design or implement effective STEAM teaching (Henriksen, 2014; Herro & Quigley, 2016a, 2016b). This relegates teachers to use existing STEM models approaching the arts or humanities as an "add-on" experience (Kim & Park, 2012; Quigley & Herro, 2016). As a result, the programs are not significantly different from current STEM education practices (Guyotte, Sochacka, Costantino, Walther, & Kellam, 2015).

In this chapter, we argue the difference between STEM and STEAM is the transdisciplinarity approach through the use of social practice theory. However, without specific examples of what this looks like in vivo, educators continue to struggle to enact meaningful STEAM practices. To address this issue, the authors developed a conceptual model of STEAM educational practices and an observation rubric intended to assess teachers' implementation of these practices during the course of a 3-year study (Quigley et al. 2017). Using this model and the longitudinal data, we focused on particular practices that contributed to students' problem-solving relevant issues but also on a practice that teachers struggled the most to conceptualize and implement, the practice of transdisciplinary teaching. This chapter attempts to define transdisciplinarity in the context of STEAM and describe the implementation successes and challenges in a variety of educational contexts.

C. F. Quigley (✉) · D. Herro · A. Baker
Clemson University, Clemson, SC, USA
e-mail: cquigley@pitt.edu

© Springer Nature Switzerland AG 2019
M. S. Khine, S. Areepattamannil (eds.), *STEAM Education*,
https://doi.org/10.1007/978-3-030-04003-1_8

Theoretical Framework

Yackman (2007) is often named as the early pioneer for developing the first framework for STEAM learning. She advocated for understanding science by understanding technology. Further, she argued that arts are crucial toward interpreting science and technology. She proposed an integrative framework wherein all disciplines are integrated but in a manner that privileges science and technology over engineering and art and then suggests connecting the disciplines through mathematics. The authors are cautious in fully adopting this work in practice because of the heavy reliance on art to *inform* the disciplines instead of as an *integral component of* problem-solving. With this view, art becomes an afterthought. Another criticism is the ever-present focus on math and science (Nanni-Messegee & Murphy 2013). Also absent in Yackman's STEAM framework is a theoretical framework to understand its conceptual grounding. Without this important framing, it is impossible to situate the work in the broader contexts of educational research and theory. As such, over the course of 3 years, we have followed STEAM education research and attempted to construct a theory-based STEAM conceptual framework, which we tested and modified based on teachers' STEAM implementation practices. One particular theory has bound the instructional practices together. This theory is called social practice theory (Roth & McGinn, 1998).

Social Practice Theory

Both STEM fields and the art fields have conceived their work as social practice. For example, Dewey's (1934) *Art as Experience* called for art to be not only a component of life but also for improving life. Similarly, STEM fields have argued that the goal of advancements through these fields should be to improve life for all. However, the way in which these social practices play out in K-12 settings is less clear. Roth and McGinn (1998) describe social practice in education settings as shared, developed, and negotiated within specific communities of knowing (Bowen, Roth, & McGinn, 1999). However, the art world expands this view point and considers, "engaging with or collaborating with a public, working across a variety of disciplines, and instigating works that have relevance to both an art and a variety of non-art audience" Guyotte et al., (2014). In this way, the key components of social practice theory are collaboration outside of school setting, discipline integration, and relevance across fields. Science and art education are not the only fields that are incorporating the theory of social practices into educational practices. Both technology and engineering education have been undergoing a reform movement that examines how the profession can contribute to creating a more just society (Bailee & Catalano, 2009). Educators are examining how engineering might look if conducted as a social practice. Hence, the goals of engineering could include ecological and social justice. By altering the conceptions of engineering and by incorporating the context for the problems that are solved, social practice theory in engineering would include examining problem-solving from socially situated context. In this fashion, we connect social practice theory as a way to understand the purpose of STEAM education. From our perspective, the outcomes of problem-solving must be socially situated, and it is critical that students are able to have opportunities to examine these problems in situ and understand their importance across fields.

Understanding the Current Field of STEAM Education

To understand the worth of implementing a transdisciplinary approach in the context of STEAM, it is helpful to differentiate transdisciplinarity from multidisciplinary and interdisciplinary, as these ideas are often conflated which creates a misunderstanding when educators attempt to implement these practices (Kaufman, Moss, & Osborn 2003). Mallon and Burnton (2005) argue multidisciplinary teaching and learning K-12 happen when experts across disciplines work "independently on different aspects of a project (p. 2)." Additionally, others understand multidisciplinarity as occurring when experts work parallel to one another while still remaining within their own disciplines (Slatin, Galizzi, Melillo, Mawn, & Phase in Healthcare Team, 2004). Interdisciplinary-structured teaching and learning builds upon multidisciplinarity, by claiming it intends to "unify two or more disciplines or to create a new 'interdisciplinary' (hybrid) discipline at the interface of the mother disciplines" (Schummer, 2004, p. 11). Finally, Nicolescu, one of the key promoters of transdisciplinary education, claims transdisciplinarity is said to occur when "knowledge corresponds to an in vivo knowledge…and includes a system of values, the humanistic values" (Nicolescu and Ertas 2013, p. 18). Many scholars agree transdisciplinary education is a holistic approach to education (Collin, 2009; Lattuca, 2003; Slatin et al., 2004); however, Arthur, Hall, and Lawrence (1989) claim it is grounded in *one* discipline while acknowledging the different viewpoints, assumptions, and findings of others. We agree with Arthur and colleagues that it often produces new perspectives but disagree that the grounding is only in one discipline. From our conceptualization, transdisciplinary teaching involves multiple disciplines where there are naturally occurring overlapped spaces between the disciplines to produce new perspectives (Gibbs, 2015). This type of problem-solving helps learners see the connections between their content and others (Pohl, 2005). When addressing teaching, Wang et al. (2011) contend that transdisciplinarity requires teachers to be able to integrate context while combining a multidisciplinary approach to blending disciplines.

What makes transdisciplinarity important for problem-solving is that it focuses on the content of one discipline and uses contexts from a different discipline to make the content more relevant. For example, a teacher might create a unit around the appropriate enclosure sizes for zoo animals. The math content would be ratios and calculating area and/or volume; however, understanding animal behavior makes the topic more relevant and would provide a better platform for solving the problem.

We view transdisciplinary inquiry as incorporating both context and content integration. As teachers design STEAM practices, the goal is to teach transdisciplinarity; however, we realize this goal may not always be attainable. That said, using any level of discipline integration provides an opportunity for multiple contents and methods to solve problems.

Today's youth will be confronted with challenges and questions that require global-view thinking to solve. The types of questions they will solve are deep-seeded, transdisciplinary issues which force comprehensive approaches to solving (Galliot, Greens, Seddon, Wilson, & Woodham, 2011). This requires a high level of creativity and is one of the reasons that creativity is one of the critical skills of the twenty-first century (Liao, 2016; Trilling & Fadel, 2009). This focus on creativity has led the push for STEAM education, and advocates of STEAM education believe that STEAM offers educators the chance to challenge their students to be creative and effective problem-solvers in today's competitive culture. Most researchers agree a truly transdisciplinary space for STEAM education should allow for each discipline within STEAM to occur in concert with one another, making it nearly impossible for students to categorize their learning into discrete disciplines (Liao, 2016). This type of authentic integration of disciplines is what the authors look for in well-designed STEAM scenarios; transdisciplinarity of STEAM education is said to have the "potentiality to address contemporary social issues, perhaps even on a global scale" (Ahn, 2015; Guyotte et al., 2014; Liao, 2016).

The novelty of introducing art into the STEM curricula has been well-received by numerous researchers and predicted to "move the needle" in transdisciplinary education (Bequette & Bequette, 2012; Liao, 2016; Maeda, 2013; Watson, 2015). Creative problem-solving approaches through artmaking and problem- or project-based learning open a new avenue for students to draw connections among their knowledge, skills, and abilities and how to use these connections in advancing their own education (Liao, 2016). By allowing students to explore and develop their own knowledge in this manner, educators hope this will provide ample opportunity for organic and self-directed innovation in teaching and learning, contrary to the typical thought of economic innovation (Land, 2013), which normally involves producing a product and ultimately a profit.

This chapter aims to look at ways educators can rethink problem-solving approaches in the classroom. Based on the authors' research and experiences, the type of problem-solving skills that best fosters real-world problems is transdisciplinary or authentic problems, which require multidiscipline thinking to solve. In fact, most scholars agree that best preparing our students for future careers must involve thinking across discipline boundaries (Berry et al., 2004; Stepien & Gallagher, 1993). Further, this type of teaching and learning can foster understanding of STEAM concepts in their real-world applications; as we know, real-world problems are typically interdisciplinary by their very nature (Asghar, Ellington, Rice, Johnson, & Prime, 2012). One approach to effective STEAM education is relevant problem-based curricula. In a problem-based learning environment, salient STEAM concepts are naturally nested in concepts of real-world problems. Generally, the problem-based approach to teaching STEAM tries to mirror the practices used by real experts to solve real-life problems within their respective fields (Crawford, 2000; Colliver, 2000). This also underscores the importance of inviting community experts into the classroom as it offers students insight into their personal experience and challenges they may face on the job.

Not only do these experts offer their first-hand experience and knowledge surrounding the context, they also offer students a tangible example for how their learning will reach beyond the walls of their classroom (Vernon, 1995). Problem-based learning (PBL) offers students connections and relevance for their learning and has proven to increase student motivation for learning (Galand, Bourgeois, & Frenay, 2005; Norman & Schmidt, 1992; Vernon & Blake, 1993; Wood, 2003) and develops a sense of importance for "responsible, professional attitudes with teamwork values" (Barrows, 1996). Some recognize the need and importance for incorporating experts into the classroom but face obstacles of local experts having difficulty finding the time to physically visit the classroom. Recently, educators have been utilizing technology to overcome and create ways their local experts can connect virtually with their students using video chat and other technology (Poulson, 2014).

STEAM education also has its critics. One of the major areas of criticism is with the amount of collaboration that this pedagogy requires. Due to STEAM education being so new, teaching resources, professional development opportunities, and even trainings are difficult for faculty to come by. Some claim there is interest in the idea of STEAM, but when it comes down to implementation, educators are easily deterred due to the vague conceptualization (Bequette & Bequette, 2012). Some also question the possibility of being able to truly pay tribute to all subjects equally without "watering down" the main purpose of STEM education (Jolly, 2014). Finally, there is also the fear that educators will incorporate "art" into a STEM curriculum, just for the sake of incorporating art into the lesson (Gettings, 2016). We understand these critiques and agree that without intentionality, the addition of the arts seems more like an afterthought inside of an integral part of the problem-solving process (Quigley, Harrington, & Herro, 2017).

The authors purport that it is the transdisciplinary approaches in the context of STEAM education that offer students the holistic and problem-based learning opportunities they need to be successful in their respective future careers. In this way, these educational practices are thought to provide an authentic method of subject integration versus simply adding in all subjects together into one lesson for the sake of doing so.

Conceptualizing Transdisciplinary STEAM Education

The goal of this chapter is to provide examples of in vivo STEAM education so teachers and teacher educators interested in STEAM-based education have research-based examples of how this transdisciplinarity practice looks in classrooms. These examples were created by examining them through our STEAM education model. Quigley et al. (2017) developed this model after several years of STEAM work with middle school teachers (Herro & Quigley, 2017; Quigley & Herro, 2016; Quigley et al., 2016).

148 C. F. Quigley et al.

From our prior studies (years 1 and 2), which included 43 teachers from 14 middle schools, we found effective STEAM teaching should position teachers to create transdisciplinarity problem-solving scenarios foregrounding problems for students to solve, using creative and collaborative skills that encompass various disciplines. This is significantly different from the beginning with the content and having students solve narrow problems (Herro & Quigley, 2016a). To illustrate the difference between the two approaches, we provide the examples below:

Transdisciplinary STEAM scenario

In May 2016, 35-yr-old Tonya was taken to the emergency room at Mary Black Hospital after complaining of a headache, some muscle pain and a fever. "Tonya" arrived with a slight fever (101°°F) and a severe case of conjunctivitis (pink eye). Doctors noted she had returned from a family vacation in Puerto Rico 3 days earlier, where she reported reading on the beach most days, eating at the hotel and local restaurants, and going on a snorkeling excursion. She had three noticeable mosquito bites. After running some blood tests to confirm their suspicions, Tonya was diagnosed with a mosquito-borne infection.

Mosquitoes are the deadliest animal on earth, leading to the death of over 1 million people each year just through transmission of malaria. Although malaria was eradicated from the US, new mosquito borne diseases such as West Nile Virus, Dengue, Zika, and Chikungunya have arrived. No vaccine or specific treatment exists for any of these illnesses. As such prevention is essential and health organizations are searching for ways to target and control problem mosquito populations.

You are member of a group working for the CDC assigned to identify Tonya's mosquito-borne illness and identify ways to control the spread and transmission of virus. In order to propose a solution, you must take numerous issues into consideration. Some of the issues include: mosquito habitat, life cycle and ecology; efficiency of virus transmission and persistence within the mosquito population; efficacy of current and new mosquito prevention technologies; ecological impacts of reducing and/or eliminating mosquitoes; and risk assessment of the case, social and economic impacts of travel bans to infected countries, and likelihood Tonya's infection may lead to an epidemic. Your proposal will be presented to the Secretary of Health and Human Services and her department. It should include evidence that multiple ideas were carefully examined to propose a solution in the best interest for the general population.

Discipline-focused teaching

Explain how mosquitos transmit diseases. Compare and contrast the diseases discussing the difference between bacteria and virus transmission including mosquito habitat, life cycle and ecology; efficiency of virus transmission and persistence within the mosquito population. Present your findings in a poster that highlights the differences between the diseases.

The differences between the approaches are (a) the STEAM teaching addresses problem-solving through a real-world application in which there is not a definite answer (e.g., the students are asked *to identify Tonya's mosquito-borne illness and identify ways to control the spread and transmission of virus*); (b) collaborative skills are required to present a solution in that the students will be placed in teams to solve the problem; and (c) multiple disciplines are acknowledged in that the scenario incorporates several disciplines. For example, engineering practices are used in determining the virus efficiency and technological advancements; English Language Arts (ELA) are addressed during the communication of evidence and persuasive essay writing during the formation of final ideas; science concepts are addressed during the investigations on viruses, transmission rates, and understanding the human body systems that are affected; technology is integrated through the use of visualization tools (e.g., Google Maps to see rates of infections) or videos (e.g., iMovie); social studies could be integrated in terms of exploring which countries are successfully battling these diseases and why (e.g., there is evidence that certain climates and geological landforms are more prone to Zika); and the students could incorporate the creative arts through creating music that evokes the feelings of contracting with a disease or writing a poem about emotions that arise during an outbreak.

This approach is sharply contrasted with the discipline-focused approach which relies heavily on science standards to have students explore the problem, with a goal of producing the same answer. One might argue the former is a more authentic teaching and learning approach as we would anticipate students encountering new questions, as they become curious about why certain people and countries are at greater risk, what the gender-bias is for certain diseases, and the technologies available to control mosquito growth. Additionally, we posit the transdisciplinary nature of the STEAM problems provides a context for creating social practices in K-12 settings.

Methodology

During our 3-year qualitative study, we determined several implementation successes and challenges STEAM teaching. We used multiple data sources including observations of teachers implementing STEAM units and teacher-designed STEAM curricula—including lesson and unit plans and teacher's reflective journals (for years 1 and 2). As the goal of this study was the transdisciplinarity component of STEAM, the data analysis focused on this aspect of STEAM.

Context

Data was collected for three years at three districts in the Southeast of US. There were seventeen schools in the study: 14 middle schools, two elementary schools. Of the three school districts, one district was in the "upstate" which had a higher SES stastus, one district is in the rural part of the state (low SES and highly diverse with a large immigrant population of largely Latino and Eastern European), and one district on the coast (racially and economically diverse). Seventy-two teachers participated in the study from across the three districts. All of the teachers participated in STEAM professional development (PD). While the specifics of the PD depended on the needs of the district, essentially, each teacher underwent at least a 1-week intensive STEAM training (40 h). During this PD, the teachers experienced a STEAM unit as students and then designed a STEAM unit for their classroom. The authors provided feedback on the units to ensure that the problem scenarios were relevant, problem-based, and transdisciplinary. We also ensured that standards were aligned and made suggestions for authentic assessment creation and ways to embed formative assessment. During the implementation of the STEAM units, the teachers were observed two times by the authors. These observations utilized the observation tool described below.

Observation Tool

Each teacher was observed at least two times by the research team. During the first year, the observation tool included brief descriptive information about the class (i.e., class size, grade level, content area); the purpose of the lesson, activities, and student arrangement (group work, teacher directed, etc.); and a narrative portion detailing what happened during the lesson. The narrative portion also focused on the success with STEAM practices and areas to further STEAM implementation. The authors completed the tool during the observation and conferenced with the teachers directly after the observation. During years 2 and 3, the authors refined the observation rubric to include STEAM-specific instructional approaches including discipline integration, problem-based approach, authentic tasks, inquiry-rich methods, student choice, technology integration, teacher facilitation, and assessments that were connected to the problem to be solved.

STEAM Curricula

Each teacher designed at least two units which included a daily plan, explicit description of components of STEAM (see observation rubric for specifics), standards, and community experts that will be involved.

Reflective Journal

Teachers kept a weekly, digital reflective journal throughout the STEAM unit (about 12–16 weeks) using it to discuss the STEAM practices they implemented, challenges, and successes they had with the implementation. Journal entries ranged from one paragraph to four paragraphs each week. These data were used as a primary data source to track the trajectory and frequency of implementation of the STEAM practices and to understand impediments to implementing STEAM practices.

Data Analysis

We analyzed the primary data sources (observations and reflections) using a priori codes. These a priori codes were taken from theoretical approaches discussed in the literature and noted from analyzing the pre-/post-data in the first phase of this study (see Quigley & Herro, 2016). These codes were discipline integration, problem-based approach, authentic tasks, inquiry-rich methods, student choice, technology integration, teacher facilitation, and assessments that were connected to the problem to be solved. Then, we conducted a second round of analysis to focus on the transdisciplinarity component. This included the level of relevant, problem-based approach, discipline integration, and the multiple ways to solve the problem. This allowed us to conceptualize transdisciplinarity similar to Kaufman and his colleagues but also expand on their work by attending to the relevance that engages students with social practice theory.

Vignettes: Understanding Transdisciplinarity in STEAM Contexts

As stated earlier, this study focused on one component of STEAM that throughout our prior research teachers found the most critical to the success of the STEAM units and at the same time the most challenging. Interestingly, this was true across all settings—regardless of the age of the students or subject area of the teacher. Transdisciplinarity includes three components: relevant, problem-based approach; discipline integration; and multiple ways to solve the problem. During our analysis, we found there were three components that led to either success or challenges in regard to STEAM implementation. These were *conceptualization of STEAM, relevant problem-based curricula design, and flexibility in enactment.* Overwhelming, without conceptualization, the teachers were not able to plan transdisciplinary units or implement them. However, there were cases were teachers had solid conceptualization and developed transdisciplinary units but were not able to enact these practices in their classrooms. The contexts within classroom impact implementation; thus we provide a variety of examples detailing the components mentioned above. We also highlight six examples of varying levels of conceptualization, curricular design, and enactment of STEAM transdisciplinarity.

Example 1 Embracing Flexibility in Planned Lessons

In grade 5, the teaching team developed a unit with the following STEAM problem scenario:

> Conde Nast Traveler and Travel and Leisure magazine just named Charleston, South Carolina the world's best city. This impressive ranking is attributed to the arts, dining, shopping and the rich history of this great American city. South Carolina coastal cities are among the fastest growing in the nation. While Charleston can be very proud of its ability

to attract people to visit and live in the area, this honor also comes with some consequences for the city. The area faces the challenge of growing at an average daily rate of 48 people per day. Rapid population growth can create a plethora of issues and problems like repairs for big-ticket road projects, lack of parking issues, overcrowded beaches, and the need for new schools. For example, in Baskerville School District, the district plans to add 3 new schools each year! One particular area of focus is that transportation engineers are looking for solutions to determine what is best for residents and the community including easing traffic congestion and facilitating faster commutes with safety and procedures. However, there are many other concerns. The Tri-county government is interested in learning about ways other cities have dealt with these issues and have asked for your help in deciding what is the most important issue. You and your team will research and decide which area the government should focus on (e.g. environmental issues, social services, tourism, education, traffic plans). At the end of this investigation, you will create a proposal for the government to review as well as a persuasive infomercial trying to convince them to choose your area.

In this example, the teachers planned a unit that begins with a problem for the students to solve: *to investigate the challenges of population growth on the area.* This problem integrates disciplines as all students research a variety of topics before choosing their area of focus and all students needed to create a persuasive essay prior to their infomercial. The students also utilized technology integration during their movie making process (they used iMovie in the process). This is an example of using social practice theory in that the students were attempting to solve a real-world problem, one that many coastal cities are struggling to solve.

During the investigations, the students discovered that many of these challenges were occurring because of another factor not mentioned in the problem scenario: climate change. Their county is in what is called the "low country" and is below sea level. This causes many issues with flooding, and the increased number of hurricanes due to the temperature and sea level rise of the ocean has increased the frequency and severity of the flooding.

When creating authentic STEAM problems from a transdisciplinarity perspective, one of the benefits and challenges is that students end up going down investigatory paths that are different than those the teacher intended. In this classroom, the teacher encouraged this, but she had to be flexible with her plan. One of the side effects of transdisciplinarity curricula that does not position one discipline over another—the methods used to solve the problem can be as varied as the disciplines studied. During discussions with teachers, they stated that although this changed the timeline of their project, they took a look at their yearlong pacing and realized that "impacts on the environment due to humans" would be studied later in the year. Therefore, they incorporated those standards in this unit, providing them with more time to focus on this unit. In this way, the flexibility the teachers had with the curriculum guides permitted the transdisciplinarity learning. Another deviation from the curricula occurred when several students discovered that similar to many cities across the United States, when the population increases, there is an initial tax on the healthcare industry. They discovered that this was already an issue for their area, and the students began to research "why?" Similar to when the students directed the learning about climate change, the teachers used this as an opportunity to historically investigate shortages in social services and the impact on the economy. They were able to connect this to the social studies standards related to reconstruction which state, "*Re*construction was a period of great hope, incredible change, and efforts at rebuilding. To understand Reconstruction and race relations in the United States, the student will Compare the political, economic, and social effects of Reconstruction on different populations in the South and in other regions of the United States" (South Carolina Department of Education, 2011). In this manner, the teachers used this as an opportunity to discuss what happens when cities change, as there are often issues related to acess of social services as well as issues related to racial inequity.

Through this example, we described the way in which STEAM curriculum that is transdisciplinary is problem-based, integrates multiple disciplines, and provides opportunities to solve problems in a variety of ways. Moreover, when it is situated in an authentic problem, there are natural connections between the disciplines (in this case persuasion, history and science). Through our conceptualization of STEAM as transdisciplinary, we are not looking to check off all the boxes of science, technology, engineering, arts, and mathematics. Instead, we found that disciplines that are naturally used to solve the problem provide a pathway for discipline integration instead of forcing *all* the subjects into the problem scenario.

Example 2 Stuck on the Facts

During a school-wide implementation of a STEAM unit about floods, the fourth-grade team designed the following problem scenario:

> On October 1, 2015, the Smithville area experienced a large amount of rain due to a stalled storm offshore. The area received 15–25 inches of rain within 48 hours. This caused a substantial amount of flooding and damage in our community, and forced many community members to be evacuated from their homes. The National Guard was called in to help rescue people, and our elementary was even used as a shelter-in- place during this time. Homes were not constructed to withstand a storm of this magnitude, so nearly a year later, families and community helpers are continuing to rebuild their homes and restore their property. Specifically, what type of preventative measures or steps can Smithville community members take in order to protect and reduce the impact that future? What type of impact does severe weather conditions could have on their homes and their families?

The plan includes a real-world problem to be solved, opportunities for discipline integration, and a specific yet open-ended question of the study allowing for students to investigate multiple paths. However, during the observations, we noted that the enactment of trandisciplinarity was lacking. Instead of investigating ways to reduce the impacts on communities during floods, the students were asked to "choose a type of severe weather" and then record their research on a teacher-created template that asked for specific facts such as "characteristics of the severe weather," "definition," "frequency," "The region(s) where their type of weather is most prevalent," and "Identify appropriate tools used to measure data, for example: anemometer, rain gauge, wind vane, or thermometer." The students were asked to keep track of their references and then create a brochure about their severe weather type.

While creating an educational brochure undoubtedly involves other disciplines (ELA to research and write, technology to create a digital brochure, science to learn

154 C. F. Quigley et al.

about the weather), the connection to solving the problem was missing. It can be argued that these brochures could help people become aware of the weather and therefore prepare for it; however, the types of facts that they were asked to include were narrow and specific. The brochures would likely look very similar across all the severe weather types. More so, this problem did not encourage multiple ways to solve the problem—in fact, we would argue the students did not solve the problem of *"what type of preventative measures or steps can Huger community members take in order to protect and reduce the impact that future,"* instead they created a list of facts about weather types.

Unlike the previous example, the students did not investigate unintended paths. In this way, the curriculum was teacher directed and did not foster opportunities for students to follow their interest. While this problem was initially situated within social practice theory in that this is a problem that has implications for the students' lives and is connected to the social context that they live, the *way* the problem scenario was enacted prohibited the implementation of this social practice theory into action. Interestingly, in this setting, the teacher had the support of the principal in flexibility. The entire school was investigating this STEAM scenario, and there were opportunities to rework the pacing guides. In fact, the principal requested that the pacing guides shift to meet the needs of STEAM education. Yet, the teacher in this example still felt a need to be in control of the curricula.

Many STEAM teachers discuss time as a major challenge. One teacher put it well when she described her middle school math classroom, "I am so impressed with the different directions the students are taking this project. I am excited about their creativity and to see how their individual strengths and interests are highlighted in their work. However, it is still really hard for me to loosen up on my plan. I am getting better but the pressure of 'keeping up' with the pacing is really hard to let go. Even when I know we are covering enough standards and doing real problem solving. It is just hard to let go." In this quote, you can feel the tension between authentic learning, student engagement, and multiple paths with the timelines of the pacing guides. This suggests that even with school supports, teachers need specific strategies to become better at facilitating learning and becoming flexible in their teaching plans.

Example 3 When Pacing Becomes the Focus

The eighth-grade math and English teachers co-planned a unit that connected their disciplines with a scenario that was locally relevant and an issue that had occurred in their city. The problem scenario was:

> The Melville Chamber of Commerce is holding a contest to help aspiring business-owners start their own restaurant. There is an empty store front in downtown Spartanburg and they are looking for the perfect restaurant to fill the space. This first round of the contest will be judged on the menu and marketability. One demographic that they are interested in tapping into is the youth of Melville. They have asked the 8th graders at Northeast middle school to be a part of this contest! With your group, your task is to create a menu inclusive of costs, and to design a marketing plan for your restaurant.

Interestingly, the teaching team held a strong conceptualization of STEAM, and during their reflection journal, the math teacher noted that, "For me, the difference of STEAM teaching is the connection to the real-world. How can I make the curricula relevant to the students' lives? And then what are the disciplines that make sense to solve the problem. When we designed the unit, it made sense that math and ELA would go well together." The ELA teacher held similar conceptions, "For us, it was easy to come up with a real-world problem that the students would need to use both math and English skills to solve." However, during the implementation, the team struggled to support the students in finishing the project. The ELA teacher described some of these challenges, "After we introduced the problem scenario, things sort of fell apart. It was hard for us keep pace together. The students were finished with their part in my class but the math component took longer. As this project relied heavily on the math, we had to wait for them to finish in math. I think this caused it to feel less real-world and the engagement of the students waned. Pacing was the major issue for us."

Pacing was a common issue across grade levels. Specifically, aligning the curricula within their schools yearlong pacing guides was often missing. Without situating the units into the pacing guide, teachers often grew concerned when units took longer than planned. When teachers did utilize long-range planning, they were often more confident about the number of standards they would cover and could alter the timeline of other units if the STEAM unit took longer than planned.

In our research, we found that when first-time teachers implemented a unit, it often took longer than planned. We discovered this occurred for several reasons. First, teachers did not incorporate enough "check points" or opportunities for teachers or peers to provide feedback on the progress of the problem-solving making the project goals difficult for students to meet. Second, during the first implementation, teachers underestimate the amount of skills that students need to support in problem-solving. These may be content-specific skills but also "soft skills" such as collaboration. Teachers found they needed to provide students with opportunities to practice these skills. Third, as STEAM units encourage teacher facilitation, this poses a challenge for teachers as they often have a specific idea of what the final product looks like. When the students move in a different direction, it can be difficult for teachers accept that final products can look different. Despite this challenge at the first implementation, we found that by the second or third implementation of a STEAM unit, the teachers were able to solve the pacing issue. Several things helped them to do this including opportunities for collaborative planning and flexibility with the pacing guides; these greatly improved that success of the transdisciplinarity STEAM implementation.

Example 4 Focus on the Final Product

The art teacher from a middle school designed this problem scenario:

For many years, Hampton Middle School has struggled with getting our students, parents, and the local community-at-large involved in school events. This year, we aim to begin solving that problem by organizing an event that everyone can participate in and enjoy. Therefore, we are creating a Hampton School Arts Fair! Students will be asked to create a craft to sell at the shop.

In this example, while the scenario began with a problem to be solved, the teacher solved the problem before the students began the process. The students might be able to choose the type of art that would increase the likelihood of their parents' participation in the event, but this would be a side effect of how the teacher solved the problem instead of allowing students to investigate why the community feels disconnected from the school. To improve this unit, the teacher might involve students in interviewing their parents and community members to understand why these events are not well attended. This may foster learning about certain issues that were keeping families from attending (schedules, language barriers, cultural considerations around "crafts").

This teacher struggled with conceptualization, a problem that would allow students to take multiple paths. Instead, this was a project that she wanted the students to complete. We often noted that teachers were caught up in projects that students were to complete, instead of involving them in a process to create the project. Here, the focus was on the goods to sell at the shop. While well intended, as the Art Fair is a fund-raising for the school and provides a community space for some of the school, it offered little transdisciplinarity.

Additionally, because the students are not solving a particular problem, it is hard to see how disciplines will be authentically integrated. Similar to the second example, the teacher argued there would be math integration in cost calculation of the goods created, but this was not explicit in the curricula, nor is it required to solve the problem. During the implementation of the unit, students created holiday goods to be sold. The connection to the problem of engaging the community was lost, and there was little evidence that students understood that their goods were to help engage the community in the event. With regard to transdisciplinarity of STEAM, it lacks a problem to be solved by the students, authentic discipline integration, and multiple pathways to solve the problem. In this example, we posit that the teacher's lack of conceptualizing STEAM created difficulty with designing the curricula and implementing a transdisciplinarity STEAM unit.

That said, typically in our research, we found that art teachers are able to design authentically situated problems with art at the center of solving the problem. We will discuss an example of this in the last example.

Example 5 Supporting Early Elementary Students Through Teacher Facilitation

In a kindergarten classroom, the teacher designed a unit based around the connection between science, social studies, and music. The problem scenario was:

The average person generates 4.3 pounds of trash per day. This is 1.6 pounds more than most produced back in 1960. Where does it all go? Approximately 55% of 220 million tons of waste generated each year in the United States ends up in one of the over 3,500 landfills. At Stone Creek elementary, we are really concerned about the amount of garbage in our area and so we recycle. But what if there are ways to reuse the materials too? Our principal

8 Moving Toward Transdisciplinary Instruction: A Longitudinal Examination...   157

has suggested that we create a play space and has asked that the Kindergarten class help design it. Can you think of ways we can turn the trash into toys?

This was the second STEAM unit the kindergarten teaching team had developed, and in this unit, they worked to ensure that the students were able to practice problem-solving. One of the challenges they noted in their previous implementation was that there were few opportunities for students to solve the problem. This problem scenario was designed with the opportunity to solve the problem in different ways. They allowed the students to brainstorm, and then, as a class, investigated the different options together. This is one of the differences between early elementary and upper elementary and middle school. The teachers found that while all the students should be given the opportunity to think about solving the problem in unique ways, in order to support the students during the inquiry phase, there needs to be more guidance. For example, once the brainstorming was complete, the class agreed that creating musical instruments was the best way to use the most trash from the school (water bottles, boxes, paper, straws, rubber bands, etc.). The students were encouraged to design their own instruments, create songs, and record their music; however, the notion of student directedness and teacher facilitation looks different in early elementary classrooms. Young students should have opportunities of choice and voice; however, they need guidance in solving the problems, and the teacher noted one way to support the students in this learning was to limit the types of pathways the students took. The authors feel that this is an important distinction to make with STEAM teaching—that across grade levels, the role of the teacher will change according to the content and needs of students.

Example 6 Strong conceptualization, transdisciplinarity design with an arts focus, successful implementation

The sixth-grade art teacher at a STEAM-focused school designed a unit wherein art moved beyond art as creativity but assisted in problem-solving. The problem scenario was:

Each year thousands of hatchling turtles emerge from their nests along the southeast U.S. coast and enter the Atlantic Ocean. Sadly, only an estimated one in 1,000 to 10,000 will survive to adulthood. The natural obstacles faced by young and adult sea turtles are staggering, but it is the increasing threats are causing them to be very close to extinction. Today, all sea turtles found in U.S. waters are federally listed as endangered, except for the loggerhead, which is listed as threatened. The XX Zoo would like to create an educational tool that will be displayed on World Oceans Day next to a student created giant sea turtle that will help visitors learn about this important species and understand the risks that sea turtles face and how they can help. When researching the migration patterns, discuss what the sea turtle is going through. Can you imagine moving from one location to another, leaving loved ones behind?

Have you ever had to go through a "migration" (i.e., life change, new situation, new school, new house)?

In this example, the teacher had strong conceptual understanding of transdisciplinarity and often discussed the importance of having art as expression as a component of the problem-solving stating that, "when student uses art as an actual part of the problem solving, it changes the way they would typically solve that problem. In the sea turtles project, once the students had thought about solving about the problem in relation to arts, their ideas changed." In order to this, the teacher had the student construct a migration relief (an art technique involving layering paper in various sizes), first of the sea turtles' migration, and then she had the students construct a "feelings" relief, describing a time when they "migrated." She found that students needed to personally connect to the migrating turtles, and trials the turtles face, with trials in their own lives. The individual shapes on the reliefs were reflective of their feelings during personal experiences with movement. The students were not simply called upon to relate to feelings but were asked to dig deeper and specify an event in which they had experienced a movement. Some students drew on feelings of moving to middle school from elementary school, moving from one church to another, or moving from one state to another when their parents experienced economic job relocation. Ultimately, they understood that movement is a necessity among all species—including humans. The students reflected on the project afterward by completing artist's statements. The artist statements confirmed that they could respond to their emotion aesthetically and that they could authentically connect with creativity based on core content.

Implications

In this section, we will discuss the implications for teachers and teacher educators who are attempting to utilize transdisciplinarity in STEAM. We frame these implications by discussing three stages of transdisciplinarity: conceptualization, curriculum design, and implementation strategies.

Conceptualization

Being able to conceptualize a new educational practice is a key component to successful implementation of that practice (Herro & Quigley, 2016a). Despite some background in or STEAM training by at least half of the participants before the PD, most had limited understanding of STEAM including transdisciplinary approaches. They viewed STEAM as addressing, but not necessarily integrating, multiple disciplines. This often led to the "ticking off the individual disciplines" as one assistant principal noted. We found that this is consistent with Son et al. (2012), suggesting teachers may understand core concepts of STEAM but struggle to clearly articulate it in theory, much less enactment. While many teachers had a conceptual understanding of how to include the arts and humanities (as part of transdisciplinary teaching), they primarily considered media arts focusing on creative ways to deliver presentations. This points to the need to have teacher educators involve arts and humanities experts in PD efforts and in the curricular design process. The teachers had difficulty moving from inter- or multidisciplinary teaching toward transdisciplinarity as a way to frame the problem-solving. That said, one strategy that assisted in better conceptualization of transdisciplinarity was collaboration. Specifically, the teachers discussed two ways in which collaboration facilitated a move toward transdisciplinary thinking. First, they believed in collaboration by incorporating other disciplines into their teaching (e.g., science teachers considering mathematical concepts). Second, collaboration helped them identify areas where they would need content expertise outside of their specific discipline. The authors agree with this conception and feel this is a way that it can be connected to social practice theory. Because STEAM requires teachers to incorporate multiple content areas, the teachers felt this type of collaboration provided them with the necessary support to incorporate multiple content areas and modes of inquiry.

Curricular Design

The other area that led to the success or difficulty of transdisciplinarity STEAM teaching was the curricular design component. While all teachers developed a STEAM problem scenario, the levels of incorporation of relevance to the students' lives and the degree to which it was problem-based varied. As noted in Example 4, the Art Fair, we noted that at times, teachers had a product in mind that they wanted to tweak to make "STEAM-like." We found this was often very difficult to do, as it would likely be irrelevant to the students' lives or be overly focused on a product making it difficult to add in a problem-solving component.

While the Art Fair example is extreme, we found that for teachers who had solid conceptualization of transdisciplinarity in STEAM, they were often fixated with doing a specific project. For these teachers, we often asked them to refer back to their standards, their long-range pacing guide to map out the breadth of the concepts students should learn in their class. With a breadth of topics in mind, we were able to help them to see the connections to the real world. Not surprisingly, this was often the most difficult part of the curriculum design process. Several supports increased the teachers' ability to do this. The first was time dedicated during the PD to allow the teachers to brainstorm, draft a problem scenario, and collaborate with their colleagues. Teachers note that during the school year, there is very little time to draft these innovative STEAM problem scenarios. The second support was feedback from the authors and their peers. This feedback helped to refine the problem scenarios which often needed more explicit connection to real-world problem-solving and ensuring that there were not disciplines forced into the problem scenario that would make it inauthentic. The last support that increases the likelihood for transdisciplinarity in the curricular design was flexibility with the pacing guide. If teachers were able to move topics around in the pacing guide, they were able to create more authentic problem scenarios that involved more subject areas. We found in our work that most districts were flexible with the pacing guide; however, in one instance, a district was unwilling to allow this flexibility which restricted the teachers' ability to move beyond multidisciplinary curricular design in most cases.

Implementation

In terms of implementation, we found that the only way teachers were able to implement transdisciplinarity was through conceptualization and problem-based curricular design. Even with these two components, all teachers struggled during their first implementation. This is understandable; however, we noted that it was important to support teachers during this process. The teachers wanted to move toward transdisciplinarity and often expressed disappointment when they were unable to do so. In the first example, the teacher understood the importance of teacher facilitation and allows students to solve the problem using different methods. Comparing this to the unit described in the second example, the teacher had difficulty letting go of the necessity for students to understand facts. Being able to connect the tasks that students do in class back to the problem-solving scenario was key. However, often, teachers wanted to guide their students to ensure the content was covered. This is understandable and likely a result of the standard-based reform movements. That said, we found when teachers were given opportunities to reflect and refine their practice in vivo, they saw this as a shortcoming and were able to alter their practice. This points to the importance of reflection and also observation that is focused on STEAM-based practices.

The other issue that helped teachers achieve success was flexibility with their daily plans. As noted in Example 1, the teacher was able to see the benefit in the direction that the students were taking the problem-solving and altered her plans to support the students in this. However, there were a couple of things supporting the teacher. The first was experience. Novice teachers had difficulty in being able to be flexible with their plans; however, more experienced teachers understood the nuances of curricular planning and could alter the trajectory of these plans later to "keep on pace," as one teacher described it. The second was the support from the school. As noted during curricular design, this varied according to district but also by school. This points to the need for consistent messages from administration and specific strategies and time dedicated to help teachers think about their long-range pacing.

As noted in this chapter, the importance of art as a part of problem-solving is critical to STEAM. However, this was often challenging during the implementation process. Primarily because of teachers' conceptions of art as noted in the conceptualization section, however without the support of related arts teachers, this was often absent. Many teachers found success when they were paired with a related arts teacher who had experience in authentic integration of disciplines. However, as noted in the last example, moving beyond integration requires allowing art to guide the solving of the problem. It is our hope that time to conceptualize and design projects with related arts teachers will allow examples such as the sea turtle project to become more prevalent.

Conclusion

This chapter provides examples in a variety of settings, grade levels, and content areas offering context-rich examples of how STEAM looks in practice. Transdisciplinarity can be a difficult strategy for teachers to incorporate. However, overwhelmingly, teachers discussed its importance as a platform for authentic problem-solving for students. As one teacher described it, "I knew I was doing STEAM when the students in my class were using science, technology, social studies and so on to solve the problem. But they were doing it on their own terms. It wasn't like they said 'oh, I am using science now' but it is what made sense to solve the problem."

References

Ahn, C. (2015). EcoScience+ art initiative: Designing a new paradigm for college education, scholarship, and service. *The STEAM Journal, 2*(1), 11.

Arthur, M. B., Hall, D. T., & Lawrence, B. S. (Eds.). (1989). *Handbook of career theory.* New York: Cambridge University Press.

Asghar, A., Ellington, R., Rice, E., Johnson, F., & Prime, G. M. (2012). Supporting STEM education in secondary science contexts. *Interdisciplinary Journal of Problem-Based Learning, 6*(2), 4.

Baillee, C., & Catalano, G. (2009). *Engineering and society: Working towards social justice [synthesis lectures on engineers, technology and society series].* San Rafael, CA: Morgan & Claypool.

Barrows, H. S. (1996). Problem-based learning in medicine and beyond: A brief overview. *New Directions for Teaching and Learning, 1996*(68), 3–12.

Bequette, J. W., & Bequette, M. B. (2012). A place for art and design education in the STEM conversation. *Art Education, 65*(2), 40–47.

Bernstein, J. H. (2015). Transdisciplinarity: A review of its origins, development, and current issues. *Journal of Research Practice, 11*(1), 1.

Berry III, R., Reed, P., Ritz, J., Lin, C., Hsiung, S., & Frazier, W. (2004). STEM initiatives: Stimulating students to improve science and mathematics achievement. *The Technology Teacher, 64*(4), 23–30.

Bowen, G. M., Roth, W. M., & McGinn, M. K. (1999). Interpretations of graphs by university biology students and practicing scientists: Toward a social practice view of scientific representation practices. *Journal of Research in Science Teaching, 36*(9), 1020–1043.

Checkland, P. (2002). *Systems thinking, systems practice; soft systems methodology: A 30-year retrospective.* Chichester, UK: Wiley.

Connor, A. M., Karmokar, S., & Whittington, C. (2015). From STEM to STEAM: Strategies for enhancing engineering & technology education. *International Journal of Engineering Pedagogy, 5*(2), 37-47.

Cnor, A. M., Karmokar, S., & Whittington, C. (2015). From STEM to STEAM: Strategies for enhancing engineering & technology education. *International Journal of Engineering Pedagogies, 5*(2), 37–47. https://doi.org/10.3991/ijep.v5i2.4458

Collin, A. (2009). Multidisciplinary, interdisciplinary, and transdisciplinary collaboration: Implications for vocational psychology. *International Journal for Educational and Vocational Guidance, 9*(2), 101–110.

Colliver, J. A. (2000). Effectiveness of problem-based learning curricula: Research and theory. *Academic Medicine, 75*(3), 259–266.

Crawford, B. A. (2000). Embracing the essence of inquiry: New roles for science teachers. *Journal of Research in Science Teaching, 37*(9), 916–937.

Delaney, M. (2014). Schools shift from STEM to STEAM. Edtech, April 2, 1–4. http://www.edtechmagazine.com/k12/article/2014/04/schools-shift-stem-steam.

Dewey, J. (1980). Art as experience (1934). *ALA Booklist, 30*, 272.

Galand, B., Bourgeois, E., & Frenay, M. (2005). The impact of a PBL curriculum on students' motivation and self-regulation. *Cashiers de Recherche en Education et Formation, 37*, 1–13.

Galliot, A., Greens, R., Seddon, P., Wilson, M., & Woodham, J. (2011). Bridging STEM to STEAM: Trans-disciplinary research. *Centre for Research & Development, Research News, 28*, 20–23. Retrieved from http://arts.brighton.ac.uk/__data/assets/pdf_file/0006/43989/Research-News-28-on-line.pdf

Gettings, M. (2016). Putting it all together: STEAM, PBL, scientific method, and the studio habits of mind. *Art Education, 69*(4), 10–11.

Gibbs, P. (2015). Transdisciplinarity as epistemology, ontology or principles of practical judgment. In P. Gibbs (Ed.), *Transdisciplinary professional learning and practice* (pp. 151–164). London: Springer International Publishing.

Guyotte, K. W., Sochacka, N. W., Costantino, T. E., Walther, J., & Kellam, N. N. (2014). STEAM as social practice: Cultivating creativity in transdisciplinary spaces. Art Education, 67(6), 12-19.

Guyotte, K., Sochacka, N., Costantino, T., Walther, J., & Kellam, N. (2015). STEAM as social practice: Cultivating creativity in transdisciplinary spaces. *Art Education, 67*(6), 12–19.

Henriksen, D. (2014). Full STEAM ahead: Creativity in excellent STEM teaching practices. *The STEAM journal, 1*(2), 15.

Henriksen, D., DeSchryver, M., Mishra, P., & Deep-Play Research Group. (2015). Rethinking technology & creativity in the 21st century transform and transcend: Synthesis as a transdisciplinary approach to thinking and learning. *TechTrends, 59*(4), 5. https://doi.org/10.1007/s11528-015-0863-9

Herro, D., & Quigley, C. (2016a). Exploring teachers' perspectives of STEAM teaching: Implications for practice. *Professional Development in Education*, 1–23. https://doi.org/10.1080/19415257.2016.1205507.

Herro, D., & Quigley, C. (2016b). Innovating with STEAM in middle school classrooms: Remixing education. *On the Horizon, 24*(3), 190–204.

Jolly, A. (2014). STEM vs. STEAM: Do the arts belong. *Education Week, 18.*

Kaufman, D., Moss, D., Osborn, T. (2003). Beyond the boundaries: A transdisciplinary approach to learning and teaching.

Kim, Y., & Park, N. (2012). The effect of STEAM education on elementary school student's creativity improvement. In T. Kim, A. Stoica, W. Fang, T. Vasilakos, J. Villalba, K. Arnett, et al. (Eds.), *Computer applications for security, control and system engineering* (pp. 115–121). Berlin: Springer. https://doi.org/10.1007/978-3-642-35264-5_16

Land, M. H. (2013). Full STEAM ahead: The benefits of integrating the arts into STEM. *Procedia Computer Science, 20*, 547–552.

8 Moving Toward Transdisciplinary Instruction: A Longitudinal Examination...

Lattuca, L. R. (2003). Creating interdisciplinarity: Grounded definitions from college and university faculty. *History of Intellectual Culture, 3*(1), 1–20.

Liao, C. (2016). From interdisciplinary to transdisciplinary: An arts-integrated approach to STEAM education. *Art Education, 69*(6), 44–49.

Maeda, J. (2013). STEM + Art = Steam. *The STEAM Journal, 1*(1), 34.

Mallon, W. T., & Bunton, S. A. (2005). The functions of centers and institutes in academic biomedical research. *Analysis in Brief, 5*, 1–2.

Nanni-Messegee, L., & Murphy, T. B. (2013). Putting theatre arts to the test: Student performance that Goes beyond STEM and STEAM. *Inquiry: The Journal of the Virginia Community Colleges, 18*(1), 6.

Nicolescu, B., & Ertas, A. (2013). *Transdisciplinary theory and practice.* Creskill: Hampton Press.

Norman, G. R., & Schmidt, H. G. (1992). The psychological basis of problem-based learning: A review of the evidence. *Academic Medicine, 67*(9), 557–565.

Padurean, A., & Cheveresan, C. T. (2004). Transdisciplinarity in education. *EDUCA IA-PLUS Journal Plus Education, 127.*

Pohl, C. (2005). Transdisciplinary collaboration in environmental research. *Futures, 37*(10), 1159–1178.

Poulson, S. (2014). Sparking student interest in STEM by bringing industry experts into the classroom. Retrieved from http://www.newschools.org/news/sparking-student-interest-in-stem-by-bringing-industry-experts-into-the-classroom/.

Quigley, C. F., & Herro, D. (2016). "Finding the joy in the unknown": Implementation of STEAM teaching practices in middle school science and math classrooms. *Journal of Science Education and Technology, 25*(3), 410–426. https://doi.org/10.1007/s10956-016-9602-z

Quigley, C. F., Harrington, J., Herro, D. (2017a). Moving beyond just adding "A" to STEM: Arts as expression. *Science Scope.* In press.

Quigley, C. F., Herro, D., & Jamil, F. (2017b) STEAM: Conceptual model for transdisciplinary learning. *School Science and Mathematics.* In press.

Quigley, C. F., Herro, D., & Jamil, F. M. (2017). Developing a conceptual model of STEAM teaching practices. *School Science and Mathematics, 117*(1–2), 1–12. https://doi.org/10.1111/ssm.12201

Roth, W.-M., & McGinn, M. K. (1998). Inscriptions: Toward a theory of representing as social practice. *Review of Educational Research, 68*(1), 135–159.

Schummer, J. (2004). Interdisciplinary issues in nanoscale research. In D. Baird, A. Nordmann, & J. Schummer (Eds.), *Discovering the nanoscale* (pp. 9–20). Amsterdam: IOS Press.

Slatin, C., Galizzi, M., Melillo, K. D., Mawn, B., & Phase in Healthcare Team. (2004). Conducting interdisciplinary research to promote healthy and safe employment in health care: Promises and pitfalls. *Public Health Reports, 119*, 60–72.

Son, Y., Jung, S., Kwon, S., Kim, H., & Kim, D. (2012). Analysis of prospective and in-service teachers' awareness of steam convergent education. *Journal of Humanities & Social Science, 13*(1), 255–284.

South Carolina Social Department of Education. (2011). *South Carolina social studies standards.* Columbia: State or province government publication.

Stepien, W., & Gallagher, S. (1993). Problem-based learning: As authentic as it gets. *Educational Leadership, 50*, 25–25.

Trilling, B., & Fadel, C. (2009). *21st century skills: Learning for life in our times.* San Francisco: Wiley.

Vernon, D. T. (1995). Attitudes and opinions of faculty tutors about problem-based learning. *Academic Medicine, 70*(3), 216–223.

Vernon, D. T., & Blake, R. L. (1993). Does problem-based learning work? A meta-analysis of evaluative research. *Academic Medicine, 68*(7), 550–563.

Wang, H. H., Moore, T. J., Roehrig, G. H., & Park, M. S. (2011). STEM integration: Teacher perceptions and practice. *Journal of Pre-College Engineering Education Research, 1*(2), 1–13. https://doi.org/10.5703/1288284314636

Watson, A. D. (2015). Design thinking for life. *Art Education, 68*(3), 12–18.

Wood, D. F. (2003). Problem based learning. *BMJ: British Medical Journal, 326*(7384), 328.

Yackman, G. (2007). STE@M education. Retrieved from http://steamedu.com

Chapter 9
Multidisciplinary Group Composition in the STEAM Classroom

John D. Sundquist

Introduction

One of the advantages of university-level courses that focus on the intersection of STEAM fields is that students have the opportunity to uncover new connections between their own areas of research and those of their fellow students. Courses that bring together students from disparate disciplines exhibit what might appear to be an ideal scenario for them to be exposed to new ideas, perspectives, and world views that they would not encounter from students in courses that are more one-dimensional in traditional curricula. One way that instructors of STEAM courses can take advantage of such diversity of the students' academic disciplines is to implement collaborative learning projects or small group activities in which students jointly focus on task on their own. Collaborative learning allows students to interact with each other and fill in gaps in their knowledge through this interaction.

The following study focuses on the impact that the diversity of students' academic disciplines has on learning in collaborative group projects in a STEAM course. In particular, I examine aspects of group composition that play a role in affecting student's level of satisfaction with their projects and their perception of the quality of collaboration among different types of learning groups. The study was carried out in a course on "Beer and Brewing in German Culture" that explores the history, science, and culture of beer in German-speaking countries. Because of the wide variety of academic disciplines represented by students in this course, it offers the opportunity to evaluate what happens when students from the same academic field work together in small groups compared to when they collaborate with others from other areas of research. Using academic majors as a way to form homogeneous or heterogeneous groups, I analyze students' self-evaluation of their learning

---

J. D. Sundquist (✉)
Purdue University, West Lafayette, IN, USA
e-mail: jsundqui@purdue.edu

© Springer Nature Switzerland AG 2019
M. S. Khine, S. Areepattamannil (eds.), *STEAM Education*,
https://doi.org/10.1007/978-3-030-04003-1_9

experience when working with others on small group projects. The research questions for the study are as follows:

RQ1: Do heterogeneous groups of students from a variety of academic majors experience more satisfaction with collaborative learning projects than homogenous groups who share the same academic major?

RQ2: Do heterogeneous groups of students from a variety of academic majors collaborate better with each other on collaborative learning projects than homogenous groups?

In section "Background," I provide background on collaborative learning and findings that examine the effects that various group-formation techniques have on collaborative learning.

Section "Methodology" provides a description of the methodology of the study as well as a description of the course and survey that was used. In section "Results," I present the quantitative data from the student survey, while in section "Discussion," I discuss these results along with some of the qualitative data. Section "Conclusions: Summary, limitations, and future research" addresses limitations and directions for future research.

Background

In previous research on group work pedagogy, the terms used to refer to small group interactive activities in learning environments overlap much with each other. For this reason, it is necessary to specify how various terms will be applied here at the beginning of this study. The following section reviews some of the terminology and addresses relevant issues pertaining to collaborative learning, group composition, and diversity of group members in a STEAM course such as the one in this study.

Collaborative Learning

One of the most common terms in educational research that has been used in the context of group work pedagogy is *collaborative learning*. As many scholars point out, including Smith and MacGregor (1992) and Barkley, Cross, and Major (2014), collaborative learning is used as a catchall term for any activities in which students work together in small groups of at least two participants, either in face-to-face interaction in the classroom or via online activities. This term is used in a general sense to refer to any kind of joint effort among students on a specific task: members of small groups are often given a clear objective – either a specific assignment by the instructor or a more open-ended task – that allows them to explore topics as a team in order to fulfill the learning objectives of the course. Activities are meant to

9 Multidisciplinary Group Composition in the STEAM Classroom   167

shift attention away from the teacher as a source of knowledge and place the students at the intersection of learning, critical thinking, exploration, and deepening of understanding.

However, others use the term *collaborative learning* in a narrow sense to reflect its roots in social constructivist approaches to pedagogy. In such a view, the term *collaborative learning* involves the emergence of shared knowledge through the dynamic interaction of group members (Topping 1996:321). As Bruffee (1999) notes, in the context of higher education, collaborative learning often involves the exchange of ideas, opinions, and experiences between students – sometimes with the teacher as a member of the group. The knowledge gained from this interaction is meant to challenge students to debate and critical evaluation and, at times, to lead to questioning of authority (Bruffee 1999:15). This narrower understanding of collaborative learning often refers to relatively open-ended tasks with less structure that may involve competition between group members or as Johnson and Johnson (1994) note, debate, conflict, and discord (67). The aim of collaborative learning in this stricter sense of the term is, according to Barkley et al. (2014), that students often become "autonomous thinkers who are able to subject various claims (including those made by their teachers) to critical scrutiny before deciding what to believe or do" (10).

Those who advocate for this more specific definition of collaborative learning often differentiate it from *cooperative learning*. Millis and Cottell (1997) as well as Flannery (1994) point out that, unlike collaborative learning tasks, cooperative learning often involves more traditional types of structured group tasks. There is less instructor- student interaction and more positive, supportive effort among the students to work together toward a common goal. Smith (1996) highlights several characteristics of cooperative learning, including positive interdependence, individual and group accountability, development of teamwork skills, and promotive interaction (74–76).

Moreover, the role of the instructor in *cooperative learning* is often understood to be a behind-the-scenes facilitator who assigns a predesigned task and checks in on students' progress but who does not participate. As Barkley et al. (2014) point out, cooperative learning is usually the term applied to group work in the STEM fields where this type of approach with tasks that involve more structure, fact-based learning, and fewer noncontroversial topics is common.

In the current study, I will refer to the activities that students take part in as *collaborative learning*, using the umbrella term for any group activities designed for student interaction. Although many of the tasks more often fall into the specific category of *cooperative learning* in this study, I opt for the superordinate term so as not to rule out the possibility that the results could be relevant for future studies that make use of either collaborative or cooperative learning approaches. Moreover, the assignments in this study exhibit some characteristics of both approaches, and a broad application of the term brings it in line with much of the recent research on interactive group learning.

Rationale for Implementing Collaborative Learning

The benefits of implementing collaborative learning in a variety of educational contexts have been so well documented over the last two decades of research studies on its efficacy, as Barkley et al. (2014) point out, one gets the impression that is now an essential aspect of any instructional approach in most disciplines (14). Although there are many positive aspects to collaborative learning, three particular benefits are directly relevant in the context of the STEAM course "Beer and Brewing in German Culture" discussed in this study.

First, collaborative learning allows students to integrate new content knowledge in with their individual background knowledge. Small group tasks enable participants to make connections between their prior understanding of the content and new insight and revelations. As Barkley et al. (2014) put it, "what students share the ability to learn depends, to a larger extent than previously assumed, on what they already know. It is easier to learn something when one already has some background than it is to learn something completely new and unfamiliar (15–16)." Through collaborative learning, students actively seek out these new connections in a supportive environment from multiple people rather than passively absorbing information from a single, static source in a lecture format.

Secondly, collaborative learning allows students to learn from each other to fill in gaps in their understanding. Although each student might have a slightly different background or knowledge base for a group assignment, each participant benefits from sharing information and perspectives. While there might be overlapping experiences and background knowledge, there are also gaps that students can fill when they interact with each other and learn from each other. Following a social constructionist model for learning, Kiraly (2014) notes that "peers and teachers working together collaboratively are thus simultaneously creating meanings and among themselves and are also internalizing meanings individually" (36). Tudge (1992) notes the effectiveness of this learning approach, stating that, "research based on this model has indicated that social interaction between peers who bring different perspectives to bear upon a problem is a highly effective means of inducing cognitive development" (159).

A third benefit of using collaborative learning techniques is the overall positive attitude toward learning that is promoted among a wide range of students. In a meta-analysis of over 300 research studies that examine collaborative learning projects, Johnson, Johnson, & Smith (2014) note the significant gains in attitude toward learning subject matter. They point out that cooperative learning promotes positive views of the course content, the discipline in general, the instructor, and the university as a whole (9). Collaborative learning also tends to improve student motivation, increase retention rate, and promote persistence following failure (Springer, Stanne, & Donovan, 1999). In particular, these positive effects are consistently valid for students from various backgrounds. As Cuseo (1996) points out, "Cooperative learning has the potential to capitalize on the contemporary wave of student diversity … by capitalizing on the multiple, socio-cultural perspectives that can be

9 Multidisciplinary Group Composition in the STEAM Classroom 169

experienced when students from diverse backgrounds are placed in heterogeneously- formed cooperative learning groups" (24). Springer et al. (1999) find support for this claim in their analysis, noting increases in performance across gender, race, academic discipline, and the number of years that a student has been at the university.

Online Collaborative Learning and Wikis

As Barkely et al. (2014) point out, online collaborative learning projects can be just as effective as traditional face-to-face learning group efforts, as long as certain technological aspects of online tools are utilized to their fullest potential (19). In particular, wikis have been used extensively over the last two decades as a way to foster a community of learning among small group members working on collaborative projects. As Wheeler, Yeomans, and Wheeler (2008) state, wikis "enable students to collaboratively generate, mix, edit and synthesize subject-specific knowledge within a shared and openly accessible digital space. The combined knowledge of the group—dubbed 'the wisdom of the masses'—is assumed to be greater than that of the individual... (989)." Wikis are by their nature capable of supporting collaborative learning since they have social communication and collaborative writing built into their platforms. In this way, they are able to support constructivist learning, the creation of learning communities, and cognitive apprenticeship (Zheng, Niiya, & Warschauer, 2015).

Wikis in particular have potential as an effective collaborative online learning tool. This is due to a number of factors: they can be set up quickly, use open-source software, are easy to maintain, and simple to contribute to (Shih, Tseng, & Yang, 2008). However, though it is clear that wikis offer a great potential for collaboration, they are not by themselves learning tools. The effective use of wikis in the classroom requires appropriate instructional design. This includes, but is not limited to, making sure that students have the necessary digital literacy skills to contribute to a wiki and student evaluation methods that are suitable for the context in which the wiki is used (Zheng et al., 2015).

Group Composition and Group Formation

An important contributing factor to the effectiveness of collaborative learning in traditional and online contexts is *group composition* or who collaborates with whom. Related to this issue of membership is the way in which students come to be put in groups, often referred to as *group formation* – either through deliberate choices on the part of the instructor or through random assignment or by choice on the part of students. In those cases in which the instructor preselects members of each group, group formation often plays a significant role in shaping the effectiveness of the group members' interaction. As Dörnyei and Malderez (1997) point out, careful attention to group formation is essential for promoting healthy and productive group dynamics. Their study of second language learners indicates that group composition affects members' roles and status within the group, the cohesion of the group as a whole, and each member's pattern of working with others (i.e., competitively, cooperatively, or individualistically) (73–74). In other words, the choice of who works with whom has multiple effects on the performance of the group as a whole and the attitudes of each individual member.

Of particular relevance for the current study is the issue whether it is more effective in some ways to form heterogeneous or homogenous groups. In general, diversity tends to promote a richer collaborative learning environment, although there is little consensus in studies of group dynamics (Magnisalis & Demetriadis, 2011). When one considers the benefits of collaborative learning that were discussed above, namely, members learning from each other, building collective knowledge, and filling in gaps in each others' understanding, it comes as no surprise that there are a number of benefits to diversity in group composition. Kizilcec (2013) analyzed MOOCs with geographically widespread membership and determined that small groups with greater geographical diversity had a wider range of viewpoints and exhibited higher quality and quantity of collaboration. Similarly, Manske, Hecking, Hoppe, Chounta, and Werneburg (2015) examined heterogeneity in groups in technology-enhanced classrooms in STEM areas and determined that those groups with varying initial levels of performance (i.e., high- vs. low-performance scores) exhibited greater overall improvements in performance than groups whose members were all at a similar level. Earlier studies such as Sharan and Sharan (1992), Cranton (1998), and Webb, Nemer, and Zuniga (2002) provide further evidence that heterogeneity, operationalized by prior performance level, background, or ability level, is a deciding factor in the success of collaborative learning. Webb and Palincsar (1996) note that heterogeneous groups allow students better peer support as well as better cross-racial and cross-gender relations.

On the other hand, others have noted several negative effects of heterogeneous group composition. Felder, Felder, and Dietz (1998) claim that some students may feel isolated and cut off from other with whom they identify in heterogeneous groups, leading to a negative attitude or less motivation. Others such as Barkley et al. (2014) suggest that students who are assigned membership based on prior academic performance may experience fewer opportunities to learn from higher performing members or to assume leadership positions (79). In a similar way, Gijlers and De Jong (2005) found that heterogeneous grouping based on prior knowledge of the content can yield positive results, although qualitative data indicate certain pitfalls: collaboration between extremely heterogeneous groups is difficult when high-achieving students do not communicate well with low-achieving students who lack background knowledge.

In some ways, more homogenous groups may provide a solution to these problems. As Brookfield and Preskill (1999) point out, students with similar backgrounds feel comfortable working together and have an excellent starting point for communicating effectively with each other. Johnson and Johnson (1996) present a number of studies that hint at some positive effects of homogenous groups, including Hooper (1992), for instance, who found that achievement and learning efficiency were highest among homogenous high-ability groups and lower among mixed high- and low-ability groups. Underwood, Jindal, and Underwood (1994) examined gender diversity in collaborative learning, finding that single-gender groups outperformed mixed-gender groups in terms of the level of activity on task and the quantity and performance in collaborative writing.

Bekele (2006) notes that homogenous groups tend to work better together on specific goals while diverse groups tend to be more effective at creating innovative solutions.

One issue concerning group composition and heterogeneity vs. homogeneity that is directly relevant to the current study is the diversity of students' field of study in the secondary educational environment. Many studies have focused on the positive effects of heterogeneity of academic achievement levels (Webb et al., 2002), levels of knowledge (Slavin, 1987), differences in personality attributes (Bradley & Herbert, 1997), or learning styles (Alfonseca, Carro, Martín, Ortigosa, & Paredes, 2006). However, there have been relatively few studies of heterogeneity vs. homogeneity that examine whether it is more beneficial to form small groups with students from disparate disciplines or whether group members perform better on a task when they all have the same academic concentrations. Only a handful of studies have focused on this topic, examining optimal group composition based on academic discipline in the context of group formation techniques. In particular, Razmerita (2011) found that when students are provided with a choice, they tend to form groups that are homogenous in terms of their level of knowledge but heterogeneous in terms of the range of academic interests. Thus, participants feel most comfortable with those who have a similar understanding of the content but who do not necessarily share the same interests. Building off of these findings, Dascalu, Bodea, Lytras, De Pablos, and Burlacu (2014) implemented an algorithm based on particle swarm optimization for achieving an optimal group composition for online group formation in collaborative learning projects. They determined that overall satisfaction with learning increased when groups were formed that maximized the heterogeneity of disciplines among group members while maintaining homogeneity of level of knowledge. In other words, the most satisfied students were those who worked with students from a variety of other disciplines but with a similar level of familiarity with the topic.

Methodology

The current study focuses on aspects of homogeneity and heterogeneity in group composition in collaborative learning projects in a multidisciplinary STEAM course. In particular, the course "Beer and Brewing in German Culture" offers an interesting testing ground for analysis due to the diversity of students' academic interests. The following sections provide a description of the course, the collaborative learning project, type of group formation, and survey instruments used to evaluate student satisfaction.

The Course

"GER 280: Beer and Brewing in German Culture" (henceforth, GER 280) is an elective course available to students at Purdue University. The course is not required of any academic program at the university, although it is one of many options of courses that students can take to fulfill a general cultural education requirement in some of the university's colleges. GER 280 addresses the history, science, and culture of beer and brewing, focusing primarily on these topics from the perspective of German-speaking countries (e.g., Germany, Switzerland, and Austria). Although the main focus is on German brewing culture, students learn about the impact of German beer on the international market and on the American craft beer industry. The course has been taught every year since 2013 and was most recently offered during the 2016–2017 academic year.

GER 280 is a semester-long course (16 weeks) split into four main units. The first of these units examines early historical accounts of brewing and beer-like beverages, including aspects of brewing in Egyptian and Sumerian cultures as well as early European views (e.g., Greek, Roman) on alcoholic beverages. This unit also presents the first Germanic accounts of ales, mead, and beer, including drinking traditions in Anglo-Saxon England, as depicted in the heroic poem *Beowulf*, as well as Norse traditions in sagas and mythology in early Germanic literature. The second unit deals specifically with the scientific aspects of brewing, namely, factors that affect the way the ingredients in beer influence the brewing process, the steps and substeps in brewing, and scientific discoveries that helped shape our understanding of brewing science. Thirdly, students explore the development of brewing and the increase in beer's production, regulation, distribution, and consumption in the Middle Ages up to the industrial revolution in Germany. The emphasis of this unit is on major economic, political, and societal changes that affected the brewing industry as it matured to become an important part of German culture today. Lastly, students learn about proper tasting technique, ways to describe the characteristics of beer styles, and they explore regional variation of the many different German beer styles and the effects of this variation on regional and national identities.

Beer and brewing are topics at the intersection of many different scientific, engineering, technology, and humanities disciplines. For this reason, students from a wide variety of academic programs are usually attracted to this course. In the semester of this research study, there were 40 students enrolled who were studying a total of X number of academic majors. As discussed in Sundquist (2015), GER 280 covers topics relevant to the following areas of study: food science, economics, religious studies, biology, chemistry, mechanical and electrical engineering, industrial engineering, German language and literature, history, art, hospitality and tourism management, and marketing.

Wiki Projects and Group Composition

As part of their required course assignments in GER 280, students were asked to complete two separate group wiki projects. Assigning two different projects allowed for counterbalancing of the two types of the student population (homogenous vs. heterogeneous) to avoid a practice effect. In other words, some students completed the first wiki as part of a homogenous group and the second as part of a heterogeneous group, while others did this in the opposite order. Students were not informed of what type of group they were a part of.

For the first wiki, the instructor put students into groups of three or four and asked them to choose from a list of instructor-generated topics that were relevant to the readings and lectures at that point in the semester. Wiki #1 dealt with the ingredients that go into beer and the steps in the brewing process (see Appendix A). Groups were also given the opportunity to choose a topic that was not on the list, as long as they discussed it with the instructor in advance. The instructor preselected the members of each group based on their academic majors. Six of the groups were homogeneous groups in which all members shared majors or had similar areas of study, while the other six groups were heterogeneous with respect to their majors/areas of study.

For the second wiki, the instructor put students into groups again according to the same principle as for the first wiki project. The students in the six homogenous groups from the first project were split up and put into various heterogeneous groups for Wiki #2, while those who were in heterogeneous groups for Wiki #1 were put together with others from their academic major on the second project. For Wiki #2, groups were asked again to choose from topics that supplemented course content at that point during the semester, exploring this time the impact of an important inventor, invention, or scientific discovery in the history of brewing. Both types of groups were provided with a list of possible topics and given the opportunity to explore a topic that was not on the list. For both projects, students were told that they should write at least 1000 words and provide any supporting images and citations on their group's final wiki page which was made available online to the class as a whole.

The two wikis were set up through the online learning management system Blackboard Learn. The instructor first made an online blank wiki available for editing by any student who was enrolled in the class. Students could contribute to and edit their own group's wiki page, linking content to it at any time, and could also view other groups' pages throughout duration of the assignment. The final product for both projects consisted of 12 separate online pages that were devoted to each wiki and provided supplementary content to the course. All pages were linked together within the course's Blackboard site. Groups had a total of 2 weeks to complete each project. To get the students to work together initially, the instructor introduced the assignment in front of the class, presented the topic choices, and showed students a list of who was to be in each group. Over the course of three subsequent class meetings, students met in 20- to 30-min sessions to work face-to-face with each other. The wikis could be edited and written at any time, including in these collaborative meetings, although most students used the time to plan and coordinate their research together. They uploaded and edited their writing online at various points outside of class time.

To form homogenous groups, it was not always possible to find students with the exact same academic major. In those cases, it was necessary to put students in groups that shared at least some aspects of their academic disciplines. For instance, if there were only two students who whose major was "chemical engineering" but others who were studying "chemistry" or "chemical engineering technology," they were all put together. Heterogeneous groups consisted of students from a wide variety of majors (e.g., hospitality and tourism management majors together with biology and mechanical engineering majors). Appendix B provides a list of the variety of majors represented by students in the course.

The Survey

In order to measure student satisfaction with the collaborative learning experience with the two wikis in GER 280, a survey was administered after completion of each wiki.

Students could opt out of the survey if they wished to. The survey was administered online through the survey software program Qualtrics, and students were able to access the survey online any time outside of class. Although students were asked to identify their small group by its number, the survey did not ask for any other identifying information. There were two parts to the survey, namely, a section on the student's self-evaluation of his/her group's wiki experience and a section in which the student evaluated another group's wiki page. There were 12 Likert-scale questions (Strongly Disagree > Disagree > Undecided > Agree > Strongly Agree) and two required open-ended questions for the first part and five Likert-scale questions and one required open-ended question on the second part. The survey is included in Appendix C.

Survey questions were meant to elicit answers from two main areas. First, a number of the questions focused on satisfaction with the final product of the collaborative project. In particular, students evaluated their own wiki according to its readability and organization, as well as their own level of interest in the topic. The five questions about another group's wiki also focused on these same topics. The second main area of focus of the questions was meant to elicit answers regarding the quality and quantity of student collaboration and the amount of mutual learning. In particular, a number of questions focused on how well students were able to learn from each other, how well they were satisfied with the amount of student interaction, and whether they would enjoy working with the same group again. Lastly, students were also asked to provide their impressions of reading other groups' wiki pages. The results of this portion of the survey are not addressed here in this study.

The open-ended questions provided qualitative data and provided a chance for students to express their views in their own words regarding these two main areas of

9 Multidisciplinary Group Composition in the STEAM Classroom

emphasis on the survey. Students were given general guidance on what they might comment on. The survey required that they provide some written feedback before submitting.

Results

The results of the survey's Likert-scale questions are provided in the following section. In particular, data from questions that deal with students' self-evaluation with respect to satisfaction with the wiki pages are provided first, followed by the results of survey questions that focus on students' assessment of the quality and extent of collaboration in their own groups. Answers to the open-ended survey items are presented in the discussion section.

Students' Self-Evaluation of Their Group's Wiki Page

There were five questions that asked students to evaluate their level of satisfaction with aspects of their own wiki page. These questions gauge students' satisfaction with the final project (Q3), their satisfaction with their learning experience (Q6), their level of interest in the topic after completing the wiki page (Q7), and their evaluation of the wiki's writing style (Q12) as well as its structure/organization (Q13). Results are presented in Table 9.1 as average scores on a 5-point scale for each question for heterogeneous and homogeneous groups, along with the difference between the two groups from the perspective of the heterogeneous groups. The results for both Wiki #1 and #2 are combined.

The average scores on all questions are very similar for both groups. In fact, none of the questions reveal a difference between groups according to nonparametric tests for statistical significance at the .05 level of significance (e.g., two-tailed Mann-Whitney U tests for the two groups' scores on each question). The highest score for both groups was for Q6 ("I feel that I learned some new aspects about the topic that I did not know before") that also elicited the highest overall score for either group on any question (4.89 for the heterogeneous group). The largest difference in average score on a question was .16 for Q6: the homogeneous group felt that

Table 9.1 Mean of Likert scores for answers to survey questions that focus on self-evaluation of satisfaction that students have their groups' wiki pages and the differences in scores between the two groups

	Q3	Q6	Q7	Q12	Q13
Students in heterogeneous groups ($N = 35$)	4.43	4.89	4.37	4.71	4.71
Students in homogeneous groups ($N = 34$)	4.59	4.74	4.44	4.62	4.65
+/− from heterogeneous group average	**−0.16**	**0.15**	**−0.07**	**0.09**	**0.07**

176   J. D. Sundquist

they learned some new aspects of the topic slightly more than the heterogeneous group (4.59 vs. 4.43). The lowest score on any question was for Q7 that deals with the level of interest in the topic upon completion of the wiki page. Both groups scored about the same on this question (4.44 vs. 4.37), with the homogenous group scoring slightly higher than the heterogeneous group.

Students' Self-Evaluation of Their Group's Collaboration

There were seven questions that asked students to evaluate aspects of group collaboration. These questions gauge students' assessment of how much they felt they learned from fellow group members (Q2), whether the student was able to interact well with the others (Q4), whether all members of the group contributed equally (Q5), whether group members collaborated well with each other (Q8), the benefits from other students' contributions (Q9), an evaluation of the students' fellow group members' work (Q10), and whether or not the student would like to work with the same group again (Q11). Results are presented in Table 9.2 as average scores on a 5-point scale for questions for heterogeneous and homogeneous groups, along with the difference between the two groups from the perspective of the heterogeneous groups. The results for both Wiki #1 and #2 are combined.

Although there are some slightly larger differences in scores in this group of questions than on the questions about students' level of satisfaction, the average scores are quite similar between both groups. Two-tailed Mann-Whitney U tests reveal that there are no significant differences on any questions between the two groups at the 0.05 level of significance. The only question that revealed a greater difference than others is Q11, whether students would like to work together again, where the homogenous group scored higher than the heterogeneous group for a difference of 0.49 (4.32 vs. 3.83). The highest average score on a question among the heterogeneous group was 4.66 (Q5), and the lowest score for this group was 3.83 on Q11. The highest average score on a question for the heterogeneous group was 4.82, also on (Q5), while the lowest score was 3.97 for Q2. Other than the large difference on Q11, other differences include Q8 ("group members collaborated well with each other"), with the homogenous group scoring .22 higher (4.65 vs. 4.43) and Q5 ("all team members contributed equally to the group's wiki page"), again, with the homogeneous group scoring .17 higher than the heterogeneous group regarding their assessment of their fellow group members' collaboration.

Table 9.2 Mean of Likert scores for answers to survey questions that focus on self-evaluation of collaboration in groups and the differences in scores between the two groups

	Q2	Q4	Q5	Q8	Q9	Q10	Q11
Students in heterogeneous groups ($N = 35$)	4.06	4.46	4.66	4.43	4.60	4.50	3.83
Students in homogeneous groups ($N = 34$)	3.97	4.32	4.82	4.65	4.71	4.56	4.32
+/− from heterogeneous group average	**0.09**	**0.13**	**−0.17**	**−0.22**	**−0.11**	**−0.06**	**−0.49**

Discussion

The results from the survey shed light on the two research questions posed in this study. They are restated here:

RQ1: Do heterogeneous groups of students from a variety of academic majors experience more satisfaction with collaborative learning projects than homogenous groups who share the same academic major?

RQ2: Do heterogeneous groups of students from a variety of academic majors collaborate better with each other on collaborative learning projects than homogenous groups who share the same academic major?

In this section, discussion will initially focus on RQ1 and the quantitative data (Likert-scale questions) as well as qualitative data (open-ended questions) on students' self-evaluation of satisfaction with the project and the learning process. Thereafter, there will be discussion of RQ2 and the qualitative and quantitative data that pertain to this research question.

Satisfaction with the Learning Experience in Heterogeneous and Homogeneous Groups

In general, findings here on the students' level of satisfaction show no sizeable differences between heterogeneous and homogenous groups when homogeneity is operationalized as sharing the same academic major. Much of the research on collaborative learning provides evidence that more diverse groups provide students with a more satisfying learning experience. More specifically in the case of the diversity of academic disciplines, studies such as Razmerita (2011) found that when students are provided with a choice, they tend to form groups that are heterogeneous in terms of the range of academic interests but homogenous in terms of how much group members know about the topic. In a similar study, Dascalu et al. (2014) found that satisfaction with the learning process increased when groups were made up of students with a variety of interests.

Despite the lack of statistically significant differences between the heterogeneous and homogenous groups on the questions about student satisfaction, there are some interesting trends that indicate that heterogeneous groups are more satisfied with their experience than homogenous groups in some ways. Three of the five questions that deal with satisfaction had higher average scores among heterogeneous group members than the homogenous groups. In other words, when this set of questions is examined as a whole, quantitative data indicate that heterogeneous groups are slightly more satisfied with the final product of their collaborative learning experience. This includes satisfaction with the writing style (Q12), the organization and structure of the wiki page (Q13), and the overall learning experience (Q6). Moreover, the highest average score on the entire survey came from the heterogeneous group when it responded to the statement "Overall, I feel that I learned some new aspects about the topic that I did not know before." The heterogeneous group averaged a very high score of 4.91 on this question, versus 4.73 for homogenous groups, or a difference of 0.18.

This slightly higher level of satisfaction with the wiki project in heterogeneous groups is partially borne out in the qualitative data, although the students' comments exhibit some differences from the survey scores. Comments to the open-ended questions fell into two main categories: satisfaction with the experience of learning from others and satisfaction with the final version of the wiki page itself. In terms of the first type of comments, both the heterogeneous and homogenous groups were generally satisfied with how much they learned from each other, although there were more positive comments among the heterogeneous groups who used descriptive words like "good," "nice," or "great" while making reference to "learning from each other" and "liking" certain aspects of the project:

1. It was nice to build off of my group members' work (student in a heterogeneous group).
2. I liked the wiki format because I could see how my contributions fit with my teammates' work (student in a heterogeneous group).
3. It was nice to collaborate on outside research in order to gain more knowledge on a specific topic (student in a heterogeneous group).
4. I had a great time writing the wiki with my group (student in a heterogeneous group).
5. Our groups worked well together, and I definitely learned from each group member (student in a heterogeneous group).
6. I liked the wiki format because I could see how my contributions fit with my teammates' work (student in a heterogeneous group).
7. I liked branching out and meeting some of the other members of the class (student in a heterogeneous group).

Some students in homogeneous groups expressed similar positive impressions of learning from each other, although there were fewer. There were only two comments that contain expressions about positive impressions of the learning experience:

8. I liked learning new things creating this wiki page with my group because each member sought out information beyond that learned in class (student in a homogeneous group).
9. It was nice to pull info from the others in the group (student in a homogeneous group).

Some of the students in homogeneous groups indicated that they felt that their contributions were lacking compared to those of others in their group or that they wished that they had written more:

10. In the end, I felt poorly about my contribution to my group's work (student in a homogeneous group).

11. In some ways, I wish that were had gone more in depth into some of the subtopics (student in a homogeneous group).

A second type of comment specifically addresses the level of satisfaction with the finished wiki page's writing style and organization/structure. In this case, the homogenous groups had more frequent expressions of satisfaction than the heterogeneous group, although the difference is only slight. Students in the homogenous group made three comments that indicated they were pleased with aspects of the writing or organization/structure of their collaboration, while only one student in the heterogeneous groups commented positively on these aspects of their wiki page:

12. Everyone was satisfied with the final page after we had all written and revised (student in a homogeneous group).
13. We have a good finished product (student in a homogeneous group).
14. In the end I feel like we created a quality, educational wiki page that explains an interesting topic at a level that allows everyone to understand (student in a homogeneous group).
15. Everyone finished their section, and the other members sections seemed well written and informed (student in a heterogeneous group).

In sum, student comments on the free-response questions support the findings from the Likert-scale survey questions. Students in heterogeneous groups expressed more satisfaction with the learning experience, using more positive descriptions and more frequent reference to learning from others in their group. Although the homogeneous group referred more to satisfaction with the final version of the wiki page of their group with respect to its writing and organization, there were relatively few comments about these aspects overall. Much like the Likert-scale questions, these open-ended questions point to a general level of satisfaction among heterogeneous groups that is higher than among the students in homogeneous groups.

Collaboration in Heterogeneous vs. Homogeneous Groups

As discussed in the review of the literature, previous research indicates that, although there is substantial evidence for the benefits of diversity in collaborative learning groups, there are some positive aspects to homogeneous group building. In particular, Hooper (1992) found that achievement and learning efficiency were higher when group members' abilities were matched up with each other, and Underwood et al. (1994) demonstrated how the level of activity on tasks and the quantity and performance in collaborative writing was higher among single-gender groups vs. mixed-gender groups. Moreover, Bekele (2006) noted that, although heterogeneous groups perform better on creative tasks that require innovation, homogenous groups work best on specific or targeted goals.

The results of the current study on homogeneity of academic major are mixed but do provide some evidence that homogeneous group members tend to collaborate more willingly with each other. In particular, the largest difference in average scores between heterogeneous and homogeneous groups on any question from the survey was in response to Q11 ("I would like to work with the same group of students another time").

Heterogeneous groups averaged 3.82 – their lowest score on any question – while the homogeneous group averaged 4.30 or a difference of 0.48. This difference was over twice as high as for any other question. In other words, those students who had a similar academic major were much more inclined to work together again than those who worked in groups with students from a variety of academic disciplines.

Of the remaining questions in the set that focuses on the level of collaboration among students, there are not as many clear-cut differences between homogeneous and heterogeneous groups. Many of the average scores differ by only 0.10 or less. However, when all seven of the questions from this set are considered, a general trend emerges: five of the seven questions that focus on collaboration point in favor of greater interaction and an interest in working together among students in homogeneous groups. This includes Q8, for instance, which dealt with the statement "Group members collaborated well with each other," for which homogeneous student groups responded with an average score of 4.64 vs. heterogeneous groups who averaged 4.44. All together, this set of questions highlights the fact that students with similar academic majors had a generally more positive impression of collaborating on group projects than students in diverse groups.

The comments on collaboration showed that both the homogeneous and heterogeneous groups generally worked well together. However, there are several indications in the qualitative data that the homogeneous group had a more positive view of their group collaboration. The phrase "worked well together" occurred 13 times in the homogenous groups' comments versus 11 times in the heterogeneous groups' comments. Aside from the exact use of this wording, the homogeneous groups made other positive comments on their level of collaboration an additional 13 times versus only 9 times for the heterogeneous group. Some examples of the positive comments made by the homogeneous group include the following:

16. I had a very positive experience with my group members. Everyone contributed and communicated equally (student in a homogeneous group).
17. It was fun working with the group on this project. I got to meet some new people in the class (student in a homogeneous group).
18. I loved working with my group and would work with them again (student in a homogeneous group).

In terms of comments that indicate a negative experience of group collaboration or a lack of group collaboration, the two groups showed similar numbers. The homogeneous groups had six comments of this type, while the heterogeneous had seven. When individual comments are analyzed, however, it becomes clear that some of the heterogeneous groups had more serious problems with collaboration. The negative comments from the homogeneous group indicate either communication problems (three comments) or simply state that the work was done largely

9 Multidisciplinary Group Composition in the STEAM Classroom

individually (three comments). Comments relating to poor communication include the following:

19. I wish there would have been better communication (student in a homogeneous group).
20. There was a low level of communication (student in a homogeneous group).

In the heterogeneous groups, there were four comments noting that the work was done individually and no comments regarding problems communicating.

A third category of negative comment is present in the heterogeneous group: general problems with working together. This is evident in the following comments:

21. It's always difficult to work with people you don't know (student in a heterogeneous group).
22. Compared to the ease of working with my last group, this project was highly unsatisfying (student in a heterogeneous group).

These comments are striking because they are the only negative comments in the entire free response section, and they come entirely from the heterogeneous groups.

One interesting aspect of the comments on the group collaboration question is that the respondents seemed to have differing views of the meaning of group collaboration.

Some viewed a group that agreed on individual tasks and then successfully completed them, working mostly alone and communicating largely online, to be an example of effective collaboration. For others, however, this same scenario was interpreted as an example of lack of true collaboration. These two views can be seen in two comments from the homogeneous groups: "Very good collaboration. All parts were completed in a timely manner." "I would say there was little to no collaboration. We just determined who was responsible for each part and did them."

These comments appear to indicate that, for some respondents, good collaboration is equivalent to fair sharing of the work and is dependent upon each individual completing his/her assigned task in a timely manner. For others, however, this type of smoothly completed individual work does not involve real collaboration.

The different understanding of what constitutes collaboration here suggests that future surveys on the topic need to separate the idea of collaboration from the concepts of fairness and equal sharing of the work, getting along well, and actual contributions to each others' work. This could be broken up into separate questions, or, when students are asked to characterize their level of collaboration, they could also be asked to provide an example. Alternatively, a short definition of collaboration could be provided. However this issue is addressed; one interesting finding of this study is the importance of taking students' varying understanding of collaboration into account.

Words that emphasize working individually and fairly were more numerous in the heterogeneous group comments than in the homogenous group comments. There were 20 uses of the words "individual," "alone," "own," "independent," "fair," or "equal" in the homogenous group versus 28 in the heterogeneous groups.

In sum, the qualitative data corroborate the quantitative data on collaboration. Students in homogeneous groups tend to collaborate more willingly and have a positive perception of working together with other members of the group than those in heterogeneous groups. The differences between the homogeneous and heterogeneous groups were not great with regard to this category, but there were more generally more positive – and no negative – views about collaborating with members of homogeneous groups. Moreover, the free response questions shed light on the possibility that what some students understand as "collaboration" might differ from others.

Conclusions: Summary, Limitations, and Future Research

The results of this study indicate that, while there were no major differences between homogeneous and heterogeneous groups when academic discipline was used as a criterion in group formation, group composition does play a subtle role in the way that students approach collaborative learning projects. The general trends in both the quantitative and qualitative data indicate that heterogeneous groups of students with a variety of academic majors can learn much from each other and are satisfied with the learning process and final outcome of their work, but homogeneous groups tend to collaborate better with each other and have a positive impression of this student-to-student interaction.

There are some limitations to the study that should be addressed in future research on homogenous and heterogeneous grouping. First of all, although GER 280 normally attracts students from a variety of STEAM fields, this particular permutation of the course did not have as wide a spectrum as expected. There were fewer students from humanities, arts, and social science disciplines, and an overwhelming majority from natural science and engineering. For example, some of the small heterogeneous groups consisted of students from electrical engineering, mechanical engineering, and chemistry with only one student from history, rather than a more diverse group of students outside of engineering and natural sciences. Perhaps if there had been a greater variety of majors, there would have been more differences between the heterogeneous and homogeneous groups. In a similar way, there were many instances in which students in "homogeneous" groups were put together even though their majors were slightly different. For example, some homogeneous groups had to include students from both chemical engineering and chemistry. Ideally, homogeneous groups would be students from the exact same discipline. The extent to which this factor played a role is unclear, but in future research studies, it would be preferable to form homogeneous groups with students from exactly the same academic program. Thirdly, as was pointed out in section "Discussion," results of the survey questions that used the term "collaborate" revealed that some students understood this term in different ways. Perhaps by nature of the online component to the project, students understood collaboration on wikis to mean a fair distribution of the workload rather than working jointly and interacting with each other.

Future studies may use a variety of terms in Likert-scale questions or in free response prompts, providing examples of the type of collaboration that is meant.

Aspects of group formation and group composition in collaborative learning projects are important to STEAM education, since they enable instructors to make the most of the unique qualities of courses with a diverse student population. Therefore, future research may examine several underexplored issues. For instance, it may be useful to compare students' self-evaluation of their collaborative learning projects with evaluation of other students' work to see if homogeneous or heterogeneous groups respond more favorably to their own work. Moreover, it would interesting to observe how the choice of a group's project topic affects the willingness of students in homogeneous or heterogeneous groups to conduct research on that topic. In the current study, students were provided with a list of possible topics from which to choose, but if they were allowed to come up with a topic on their own, there might be increased collaboration, more learning from each other, and higher levels of satisfaction with the final product. Studies on these and similar topics about group composition, group formation, and the use of wikis would shed light on even more ways that collaborative learning in STEAM education can be effective.

Appendices

Appendix A: Wiki Project Description

GER 280: Beer and Brewing in German Culture
Group Wiki Project #1

There will be two "Wiki Projects" completed this semester in GER 280 (each worth 5% of the semester grade). You will work together in groups of three or four students (the groups will be assigned by the instructor). You will be put in groups with different students for the two assignments; you'll have several opportunities to work together in class and encouraged to meet outside of class as well. Each group's wiki page should end up being approximately 1000 words.

The goal of the project is that your group will get a chance to explore a topic that is covered in class in more depth, collaborate on research on this topic, and write up a page of information on the topic that other members of the class will be able to benefit from. Your instructor will provide the class with a list of possible topics to choose from.

Your group's wiki page will be evaluated on the following criteria:

- Informational content (relevant, thorough, and accurate information)
- Clarity of the writing (easy to understand, written for nonspecialists)
- Organization/structure of the wiki page
- Level of collaboration (that all members of the group took part in the writing)
- Proper citation methods and reference to sources

Each group member will also evaluate each other's contributions to make sure that everyone collaborated.

Here are possible topics for Wiki Project #1 (note: the description of possible sub-topics is not exhaustive nor is it a required list of items to discuss):

1. *Hops:* its purpose, history, its effects, how it fits into the brewing process, different hops-growing regions
2. *Malting:* different grains, different techniques, steps, problems that can arise, types of barley, other cereals
3. *Mashing:* its purpose, different types, effects on beer styles and taste, different styles, different types of equipment
4. *Wort:* its purpose, different variables, ways that it affects taste and style, measuring techniques, equipment
5. *Fermentation and yeast:* historical aspects, different yeast strains, how it affects the final product, problems during fermentation, chemical processes
6. *Water:* different effects of water, different minerals, variables, how they affect the styles and taste of beer, boiling
7. *Cooling:* reasons for cooling during the brewing process, ways to cool beer at different breweries, refrigeration techniques, equipment; effect of temperature during steps of the brewing process
8. *Lautering:* its purpose, variables that affect it, problems, equipment
9. *Conditioning:* different techniques, purposes, historical techniques, equipment
10. *Packaging:* different techniques, purposes, historical techniques, equipment

If your group has another idea for a project topic, please feel free to check with your instructor.

Appendix B: List of Academic Majors of Students in GER 280

Aeronautical & Astronomical Engineering
Biological Engineering
Biomedical Engineering
Chemical Engineering
Chemistry
Civil Engineering
Communication
Computer Engineering
Computer Science
Electrical Engineering
Electrical Engineering Technology
Food Science
German

9 Multidisciplinary Group Composition in the STEAM Classroom

History
Industrial Design Prof Program
Marketing
Mechanical Engineering
Mechanical Engineering Technology
Pharmacy
Physics
Public Relations and Strategic Communication

Appendix C: Survey

I. **Evaluation of Your Own Wiki Page**

Please respond to the following questions using the following scale:

1. *Strongly disagree*
2. *Disagree*
3. *Neither disagree nor agree*
4. *Agree*
5. *Strongly agree*

Q1. Your group number:_____

Q2. While working on my group's wiki page, I learned a lot from my fellow group members.

Q3. I am satisfied with the final version of my group's wiki page.

Q4. I was able to interact well with team members while we worked on the wiki page.

Q5. All team members contributed equally to the group's wiki page.

Q6. I feel that I learned some new aspects about the topic that I did not know before.

Q7. My level of interest in the topic increased after completing the wiki page.

Q8. Group members collaborated well with each other.

Q9. I benefited from other students' contributions.

Q10. I think the other team members' contributions were excellent.

Q11. I would like to work with the same group another time.

Q12. I feel the final version of our wiki page is easy to understand and written for nonspecialists.

Q13. I feel the final version of our wiki page is well-organized and easy to follow.

Q14. Comment on your learning experience in working with your group (positive and negative experiences).

Q15. Comment on the amount of collaboration with your fellow group members (note: please do not refer to specific group members).

II. Evaluation of Another Group's Wiki Page

Look over another group's wiki page on a topic different from your group's and fill out the following survey questions using the same scale:

1. *Strongly disagree*
2. *Disagree*
3. *Neither disagree nor agree*
4. *Agree*
5. *Strongly agree*

Q16. What group number's wiki page did you read? _____

Q17. Overall, the content of this wiki page's information is relevant.

Q18. The page is organized well, and the content is easy to follow.

Q19. I learned something new and interesting about the topic.

Q20. The page is easy to understand and written for nonspecialists.

Q21. Comment on any aspects of the wiki page that you read (e.g., content, organization of page, layout, accuracy, level of interest).

References

Alfonseca, E., Carro, R. M., Martín, E., Ortigosa, A., & Paredes, P. (2006). The impact of learning styles on student grouping for collaborative learning: A case study. *User Modeling and User-Adapted Interaction, 16*, 377–401.

Barkley, E. F., Cross, K. P., & Major, C. H. (2014). *Collaborative learning techniques: A handbook for college faculty*. San Francisco: Jossey-Bass/Wiley.

Bekele, R. (2006). Computer-assisted learner group formation based on personality traits. Doctoral dissertation, University of Hamburg. http://ediss.sub.uni-hamburg.de/volltexte/2006/2759/

Bradley, J. H., & Hebert, F. J. (1997). The effect of personality type on team performance. *Journal of Management Development, 16*, 337–353.

Brookfield, S. D., & Preskill, S. (1999). *Discussion as a way of teaching*. San Francisco: Jossey-Bass/Wiley.

Bruffee, K. A. (1999). *Collaborative learning: Higher education, interdependence, and the authority of knowledge*. Baltimore: Johns Hopkins University Press.

Cranton, P. (1998). *No one way: Teaching and learning in higher education*. Toronto, ON: Wall & Emerson.

Cuseo, J. B. (1996). *Cooperative learning: A pedagogy for addressing contemporary challenges & critical issues in higher education*. Stillwater, OK: New Forums Press.

Dascalu, M. I., Bodea, C. N., Lytras, M., De Pablos, P. O., & Burlacu, A. (2014). Improving e-learning communities through optimal composition of multidisciplinary learning groups. *Computers in Human Behavior, 30*, 362–371.

Dörnyei, Z., & Malderez, A. (1997). Group dynamics and foreign language teaching. *System, 25*, 65–81.

Felder, R. M., Felder, G. N., & Dietz, E. J. (1998). A longitudinal study of engineering student performance and retention. V. Comparisons with traditionally-taught students. *Journal of Engineering Education, 87*, 469–480.

9 Multidisciplinary Group Composition in the STEAM Classroom

Flannery, J. L. (1994). Teacher as co-conspirator: Knowledge and authority in collaborative learning. *New Directions for Teaching and Learning, 59*, 15–23.

Gijlers, H., & De Jong, T. (2005). The relation between prior knowledge and students' collaborative discovery learning processes. *Journal of Research in Science Teaching, 42*, 264–282.

Hooper, S. (1992). Effects of peer interaction during computer-based mathematics instruction. *The Journal of Educational Research, 85*, 180–189.

Johnson, D. W., & Johnson, R. T. (1994). Structuring academic controversy. In S. Sharan (Ed.), *Handbook of cooperative learning methods* (pp. 66–81). Westport, CT: Greenwood Press.

Johnson, D. W., & Johnson, R. T. (1996). Cooperation and the use of technology. In D. H. Jonassen (Ed.), *Handbook of research for educational communications and technology: A project of the Association for Educational Communications and Technology* (pp. 1017–1044). New York: Simon & Schuster Macmillan.

Johnson, D. W., Johnson, R. T., & Smith, K. A. (2014). Cooperative learning: Improving university instruction by basing practice on validated theory. *Journal on Excellence in University Teaching, 25*, 1–26.

Kiraly, D. (2014). *A social constructivist approach to translator education: Empowerment from theory to practice*. London: Routledge.

Kizilcec, R. F. (2013). Collaborative learning in geographically distributed and in-person groups. In Z. Pardos, & E. Schneider (Eds.) *AIED 2013 Workshops Proceedings Volume* 1 (pp. 67–74). http://ceur-ws.org/Vol-1009/aied2013ws_volume1.pdf#page=72

Magnisalis, I., & Demetriadis, S. (2011). Modeling adaptation patterns in the context of collaborative learning: Case studies of IMS-LD based implementation. In *Technology- enhanced systems and tools for collaborative learning scaffolding* (pp. 279–310). Berlin: Springer.

Manske, S., Hecking, T., Hoppe, U., Chounta, I. A., Werneburg, S. (2015). Using differences to make a difference: A Study in heterogeneity of learning groups. In *11th international conference on computer supported collaborative learning (CSCL 2015)*. https://telearn.archives-ouvertes.fr/hal-01206688/document

Millis, B. J., & Cottell. (1997). *Cooperative learning for higher education faculty*. Phoenix, AZ: Oryx Press.

Razmerita, L. (2011). Collaborative learning in heterogeneous classes: towards a group formation methodology. In *The 3rd international conference on computer supported education (CSEDU 2011)* (pp. 189–194). http://openarchive.cbs.dk/handle/10398/8335

Sharan, Y., & Sharan, S. (1992). *Expanding cooperative learning through group investigation*. New York: Teachers College Press.

Shih, W., Tseng, S., & Yang, C. (2008). Wiki-based rapid prototyping for teaching-material design in e-learning grids. *Computers & Education, 51*, 1037–1057.

Slavin, R. E. Developmental and motivational perspectives on cooperative learning: A reconciliation. *Child Development, 1987, 58*, 1161–1167.

Smith, B. L., & MacGregor, J. T. (1992). What is collaborative learning? In A. Goodsell, M. Maher, & V. Tinto (Eds.), *Collaborative learning: A sourcebook for higher education* (pp. 10–26). University Park, PA: National Center on Post-Secondary Teaching, Learning, and Assessment.

Smith, K. A. (1996). Cooperative learning: Making "groupwork" work. *New Directions for Teaching and Learning, 67*, 71–82.

Springer, L., Stanne, M. E., & Donovan, S. S. (1999). Effects of small-group learning on undergraduates in science, mathematics, engineering, and technology: A meta-analysis. *Review of Educational Research, 69*, 21–51.

Sundquist, J. D. (2015). Beer and brewing in German culture: Bridging the gaps within STEAM. *The STEAM Journal, 2*(1), 7.

Topping, K. J. (1996). The effectiveness of peer tutoring in further and higher education: A typology and review of the literature. *Higher Education, 32*, 321–345.

Tudge, J. (1992). Vygotsky, the zone of proximal development, and peer collaboration: Implications for classroom practice. In L. C. Moll (Ed.), *Vygotsky and education: Instructional implications and applications of sociohistorical psychology* (pp. 155–172). New York: Cambridge University Press.

Underwood, G., Jindal, N., & Underwood, J. (1994). Gender differences and effects of co-operation in a computer-based language task. *Educational Research, 36*, 63–74.

Webb, N. M., Nemer, K. M., & Zuniga, S. (2002). Short circuits or superconductors? Effects of group composition on high-achieving students' science assessment performance. *American Educational Research Journal, 39*, 943–989.

Webb, N. M., & Palincsar, A. S. (1996). Group processes in the classroom. In D. Berliner & R. Calfee (Eds.), *Handbook of educational psychology* (pp. 841–873). New York: Prentice Hall International.

Wheeler, S., Yeomans, P., & Wheeler, D. (2008). The good, the bad and the wiki: Evaluating student-generated content for collaborative learning. *British Journal of Educational Technology, 39*, 987–995.

Zheng, B., Niiya, M., & Warschauer, M. (2015). Wikis and collaborative learning in higher education. *Technology, Pedagogy and Education, 24*, 357–374.

Printed in the United States
By Bookmasters